AF361635

Indoor Environment Quality and Health in Energy-Efficient Buildings

Indoor Environment Quality and Health in Energy-Efficient Buildings

Editor

Roberto Alonso González Lezcano

MDPI • Basel • Beijing • Wuhan • Barcelona • Belgrade • Manchester • Tokyo • Cluj • Tianjin

Editor
Roberto Alonso González Lezcano
Departamento de Arquitectura y Diseño,
Escuela Politécnica Superior,
Universidad CEU San Pablo
Spain

Editorial Office
MDPI
St. Alban-Anlage 66
4052 Basel, Switzerland

This is a reprint of articles from the Special Issue published online in the open access journal *Sustainability* (ISSN 2071-1050) (available at: http://www.mdpi.com).

For citation purposes, cite each article independently as indicated on the article page online and as indicated below:

LastName, A.A.; LastName, B.B.; LastName, C.C. Article Title. *Journal Name* **Year**, *Volume Number*, Page Range.

ISBN 978-3-0365-3665-1 (Hbk)
ISBN 978-3-0365-3666-8 (PDF)

Contents

About the Editor

Roberto Alonso González Lezcano Tenured Professor at the Department of Architecture and Design, area of Building Systems, within the Institute of Technology of Universidad CEU San Pablo. Coordinator of the Mechanical Systems area. Professor Accredited by ANECA in the figures of Senior Professor. One six-year research period by CNEAI (period 2003–2015). Extraordinary PhD Award. Angel Herrera Award for the best research work in the field of Architecture and Engineering at the CEU San Pablo University in the XXV Edition of the Angel Herrera Awards (2021). He is Principal Investigator of the research group ARIE: Architecture, efficient installations at the Universidad CEU San Pablo. Coordinator of the post-graduate degree of Energy Efficiency and Mechanical Systems in Buildings and Coordinator of the Laboratory of Building Systems within Universidad CEU San Pablo. Member of the PhD Program in "Health Science and Technology" and the PhD Program "Composition, History and Techniques pertaining to Architecture and Urbanism"; where he is the Coordinator of the Construction, Innovation and Technology line and Coordinator of the training complement "Methodology of technical and statistical experimentation". Lecturer of the Master MC2 relevant to Quality in Construction and in the Postgraduate Degree in Overall Management of Buildings and Services at the Universidad Politécnica de Madrid. Coordinator of the Wind Energy Section within the Master's degree in Renewable Energy of the Institute of Technology (Universidad CEU San Pablo). Lecturer of Advanced Manufacturing Processes in the Master's degree in Industrial Engineering at the Universidad Europea de Madrid. In addition, lecturer of Mechanics of Continuous Media and Theory of Structures, Electromechanics and Materials in seven Spanish universities. 29 books published, 20 of them in the last 5 years. Coordinator and co-author of the collections of Abecé of Building Systems (Munilla-Lería Eds.) consisting of 5 volumes. Building Systems in Building Design (Asimétricas Eds.) consisting of 7 volumes. Articles concerning areas of energy, systems, and mechanics of continuous media (58 articles in the last 5 years; 41 JCR and 17 SJR). Editor of several books on issues of sustainability, energy efficiency and indoor environment quality indexed in Book Citation Index, WoS, Scopus, SPI among others. Director of the PhD thesis "Efficiency of ventilation in residential environments for the promotion of health of its occupants: impact on architectural indoor design" and "Energy simulation as a prognostic tool in architecture: design of passive strategies in different climatic zones of Spain" which obtained in June 2018 and September 2020, respectively, the grade of Cum Laude with mention of an International PhD. Director/Co-director of 3 PhD theses currently in progress related to energy efficiency, comfort and indoor air quality. "Influence of occupants on energy consumption in multi-family dwellings in Madrid Characterization of their habits.", "Building Systems in efficient office buildings" and "Energy simulation as a forecast tool in architecture: The importance of the solar reflectance index in the design stage of residential buildings in Spain". Member of the Editorial Board and Review Board of several journals indexed in JCR, Guest Editor of special issues and Topic Editor of journals indexed in the JCR of the MDPI publishing house.

Article

Impact of Air Infiltration on IAQ and Ventilation Efficiency in Higher Educational Classrooms in Spain

Irene Poza-Casado, Raquel Gil-Valverde, Alberto Meiss and Miguel Ángel Padilla-Marcos *

RG Architecture & Energy, Universidad de Valladolid, Avda/Salamanca, 18-47014 Valladolid, Spain; irene.poza@uva.es (I.P.-C.); raquel.gil@uva.es (R.G.-V.); meiss@arq.uva.es (A.M.)
* Correspondence: miguelangel.padilla@uva.es

Abstract: Indoor air quality (IAQ) in educational buildings is a key element of the students' well-being and academic performance. Window-opening behavior and air infiltration, generally used as the sole ventilation sources in existing educational buildings, often lead to unhealthy levels of indoor pollutants and energy waste. This paper evaluates the conditions of natural ventilation in classrooms in order to study how climate conditions affect energy waste. For that purpose, the impact of the air infiltration both on the IAQ and on the efficiency of the ventilation was evaluated in two university classrooms with natural ventilation in the Continental area of Spain. The research methodology was based on site sensors to analyze IAQ parameters such as CO_2, Total Volatile Organic Compounds (TVOC), Particulate Matter (PM), and other climate parameters for a week during the cold season. Airtightness was then assessed within the classrooms and the close built environment by means of pressurization tests, and infiltration rates were estimated. The obtained results were used to set up a Computational Fluid Dynamics (CFD) model to evaluate the age of the local air and the ventilation efficiency value. The results revealed that ventilation cannot rely only on air infiltration, and, therefore, specific controlled ventilation strategies should be implemented to improve IAQ and to avoid excessive energy loss.

Keywords: indoor air quality; thermal comfort; airtightness; natural ventilation; educational buildings

Citation: Poza-Casado, I.; Gil-Valverde, R.; Meiss, A.; Padilla-Marcos, M.Á. Impact of Air Infiltration on IAQ and Ventilation Efficiency in Higher Educational Classrooms in Spain. *Sustainability* **2021**, *13*, 6875. https://doi.org/10.3390/su13126875

Academic Editor: Vincenzo Torretta

Received: 20 May 2021
Accepted: 15 June 2021
Published: 18 June 2021

Publisher's Note: MDPI stays neutral with regard to jurisdictional claims in published maps and institutional affiliations.

1. Introduction

Indoor air quality (IAQ) is essential to create healthy and comfortable spaces for users in which to carry out their daily activities, especially considering that people spend more than 90% of their time indoors [1]. There is a large volume of published studies describing the role of IAQ in the health of building users [2,3]. When the use of the building is academic, the importance of IAQ is higher because poor indoor air for long academic periods can have a negative impact on both the students' health and intellectual performance [4,5].

The European ventilation standard (CEN) [6] establishes mechanical ventilation as the strategy to guarantee the minimum outdoor airflow to maintain adequate IAQ. In Spain, regulations established mandatory mechanical ventilation in educational buildings in 2007 [7]. A total of 90% of the compulsory educational buildings in Spain were built before 2007 when mechanical ventilation systems were not mandatory [8]. In the case of higher educational centers such as universities, the share of buildings constructed before ventilation systems were mandatory remains unknown. However, it could presumably be higher since they are often located in historical buildings. Therefore, educational buildings in Spain are mainly naturally ventilated. Ventilation (single-sided or cross ventilation) is only promoted by the occasional opening of windows when users consider it necessary, based on no objective IAQ indicator [9]. Thus, maintaining acceptable IAQ during the whole academic period remains a challenge [10]. This manual control of the openings is even more limited during the cold season since it causes uncomfortable airflows and a considerable drop in the classroom hygrothermal conditions.

In this scenario, air infiltration through cracks and other unintentional openings in the building envelope is the only mechanism that can provide the continuous renovation of the indoor air. For this reason, it is important to determine the airtightness of the building envelope, since it impacts directly on air infiltration, caused by pressures gradient [11].

This paper addresses the indoor conditions of a higher educational building with natural ventilation in the Continental climate area of Spain, considering different parameters such as temperature, relative humidity (RH), CO_2, TVOC, and PM, assessing the ventilation performance through the analysis of the ventilation efficiency caused by air infiltration.

Background

In recent years, there has been an increasing amount of literature on indoor comfort in university buildings [12–15]. Although most of the previous research focuses only on IAQ, others have considered the Indoor Environmental Quality (IEQ), focusing also on physical parameters related to sound and visual comfort as well as user's perception through questionnaires [16–18].

There is a wide range of airborne pollutants that can affect users inside buildings; among them is CO_2, which is closely connected to human activity since it is mainly caused by users' breath. Harmless as it may be considered, a high concentration of CO_2 can have a negative impact on the attention of the students, hindering academic achievement [19]. It is important to highlight that CO_2 concentration has been associated with the presence of other pollutants. Therefore, it is often considered as a good indicator of the air change capacity of the rooms and it is important to evaluate its performance avoiding unventilated areas. However, some sources warn that CO_2 should not be considered as a unique IAQ indicator [20,21]. It has been suggested that other compounds such as particulate matter (PM) or total volatile organic compounds (TVOCs) should be also considered to assess IAQ [22].

Volatile organic compounds (VOCs) can be produced by the construction materials of the building, its furniture, as well as cleaning products and academic materials such as adhesives, paints, or office machinery [15]. Common sources of PM include human activity, the use of scholarly materials, or even the outdoor environment [23]. High concentrations of these pollutants can produce irritation of the mucous membranes, fatigue, headaches, or aggravation of asthma in the users, among others [24]. Previous research has suggested that TVOC concentration could be related to sick building syndrome (SBS) [25].

The relationship between airtightness and IAQ in educational buildings in Mediterranean countries with natural ventilation has already been addressed in previous research [9,26,27]. No direct relationship between airtightness and IAQ was found in a study that assessed 42 classrooms, especially if windows were open [9]. In this regard, previous research in southern European climates has revealed that air infiltration must be complemented with other ventilation mechanisms in order to maintain good IAQ [26,27].

However, air infiltration in buildings is usually not considered as part of the design of the ventilation system [1]. This causes a non-predictive air track model that entails an inhomogeneous air quality distribution through the indoor space, whose impact depends on the occupants' location [28]. The distribution of the air quality can be evaluated by assessing the ventilation efficiency and age of the air [29], which can be modeled by knowing the air infiltration rate and the position of the air leakage paths using Computational Fluid Dynamics (CFD) software. Yet the existing literature has overlooked this approach, which relates air infiltration, the efficiency of ventilation, and IAQ to verify the suitability of air leakages as a ventilation source.

2. Materials and Methods

2.1. Site and Building

Two classrooms of the Campus "Duques de Soria" (Universidad de Valladolid), built in 2006, were assessed. The campus is located in a suburban area of Soria (Spain), which has a temperate climate type "Cfb" in the Köppen-Geiger classification. The summer is short and hot, and the winter is cold, long, and windy, both seasons being very dry. The average temperature during the summer is 23 °C and 10 °C during the winter. The wind has a predominant direction west–north–west, and its annual mean velocity is 10.8 km/h (type 2 in the Beaufort scale [30]).

The envelope of the building where the classrooms under study are located is made of sandstone boards that are fixed to 11.5 cm of perforated brick, thermal insulation of extruded polyethene, vertical air chamber, and an interior layer of 11.5 cm of double-hollow brick. The windows are made of aluminum profiles with double glazing. Rolling shutters, a typical mechanism used in Spain to provide shadow and protection, are built within a non-integrated box over the windows. The classrooms have a false ceiling 70 cm high made of acoustic panels of rigid plaster. The heating system has hot water radiators made of aluminum as terminal units.

For the purpose of the study, the most unfavorable classrooms of the building in terms of IAQ based on its configuration and use were determined. To that end, the type of ventilation, occupancy, volume, orientation, and nearby environment were considered. Concerning ventilation, all the classrooms of the building had natural ventilation, which is considered the most unfavorable scenario. The determination of the classrooms with higher levels of occupation over time was done according to the classrooms' lesson schedules, considering the number of students and the classroom's volume. Also, classrooms located near the car park were chosen due to the possible negative impact of vehicles on the outdoor air conditions. Orientation was considered, given its relationship with indoor temperature, because the heating system was homogeneously designed. Classrooms with the main façade facing south were expected to reach higher indoor temperatures and, therefore, have poorer conditions regarding IAQ. In this sense, it was first taken into account that high temperatures favor chemical reactions that lead to the production of certain pollutants such as VOCs or microorganisms [31]. Secondly, lower infiltration rates were expected as a result of a lower temperature gradient, causing a reduction of the stack effect [32] in these classrooms during the cold season.

After the analysis of the aforementioned parameters, it was determined that the block of classrooms A04 and A05 was the most unfavorable of the campus, with the following characteristics (Table 1 and Figure 1).

Table 1. Classroom characteristics.

Classroom	Floor	Orientation	Area (m^2)	Average Occupation Time	Height (m)	Volume (m^3)	Doors (m^2)	Windows (m^2)
A04	ground-floor	SE	103.7	4.2 h/day	2.96	307.6	1 (1.6 × 2)	4 (1.4 × 1)
A05	ground-floor	SE	92.6	5.8 h/day	2.96	274.6	1 (1.6 × 2)	10 (1.4 × 1)

2.2. IAQ Equipment and Parameters

Air quality measurement stations were used to register variations in the levels of different air contaminants that are usually present in the air (CO_2, TVOC, and $PM_{1.0}$, $PM_{2.5}$, and $PM_{10.0}$). These parameters were chosen as indicators because of their presence in educational environments and their influence on the students [22]. Each device registered temperature, Relative Humidity (RH), and barometric pressure as well, saving all the information in the data logger of the station.

Figure 1. Location of classrooms A04 and A05 within the building.

The devices integrated different sensors to determine the level of each parameter. Model K30 10,000 ppm was used to measure CO_2, with a margin of error of ±30 ppm/$\pm3\%$ of the measured value. TVOC concentration was measured with an iAQ-Core Indoor Air Quality Sensor Module device, whereas PM was measured with a Digital Universal particle concentration sensor PMS5003 with a margin of error of $\pm10\%$. The temperature and RH sensor SHT2x had a margin of error of $\pm0.3\,^\circ$C and $\pm1.8\%$ RH.

Concerning the temperature and RH analysis, the ranges specified by Spanish regulations (RITE [33]) for the cold season were taken into account: temperature between 21 and 23 $^\circ$C, and RH between 40 and 50%.

The CO_2 analysis was done considering the values established in the Spanish regulations [33] as well. This regulatory framework establishes CO_2 as an indicator of the human bio-effluent emissions, used to set the minimum ventilation airflow needed. RITE defines classrooms as places where good IAQ is needed and fixes a maximum corrected concentration of 500 ppm above the CO_2 concentration in the outdoor air to reach acceptable conditions. Therefore, the corrected concentration is obtained by deducting the average CO_2 concentration outdoors from the values registered inside the building. However, no consideration is made concerning TVOC or PM concentration by RITE. An acceptable range of up 500 ppb of TVOC concentration within the classroom environment was considered, according to several sources and recommendations [22,34,35]. In the case of PM, the maximum concentrations of $PM_{2.5}$ and $PM_{10.0}$ were set at 25 and 50 μg/m^3, respectively [36].

2.3. IAQ Monitoring

Measurements were carried out between February and March for a week. The data collection was done simultaneously in classrooms A04, A05, the adjacent corridor, and the exterior of the building. Measurements were taken for approximately 5 s every minute during the whole academic period, although the subsequent analysis of the data focused on times when evidence of occupation in the classroom was found, which corresponds to the highlighted parts of the schedule: this was 21 h in classroom A04 and 28 h in classroom A05 (Figure 2).

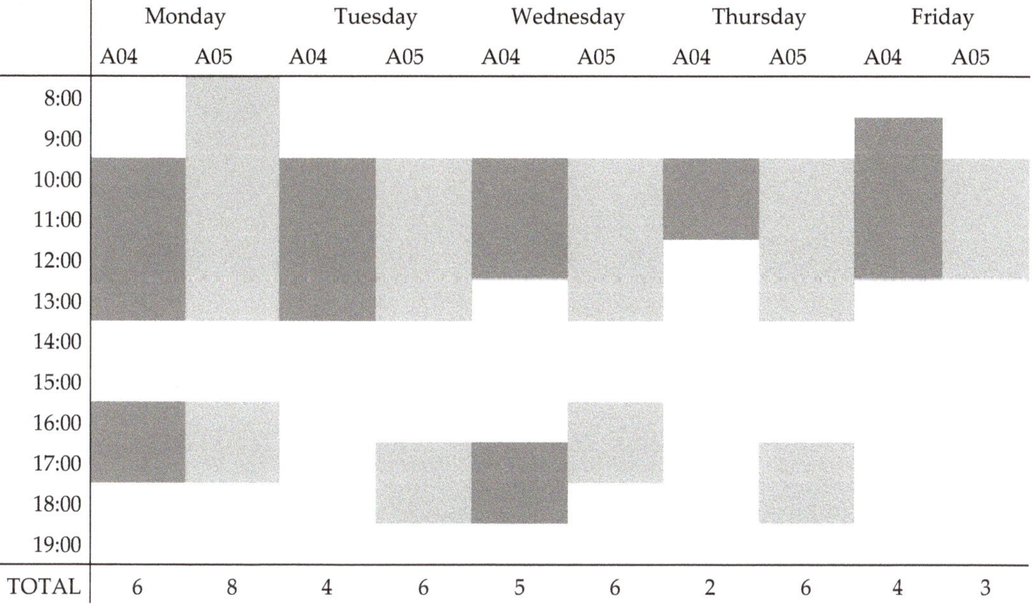

Figure 2. Classroom schedule.

The location of the devices inside the classrooms was determined to avoid distortions: separated from the usual traffic areas of the students and the teacher, and one meter away from doors or windows. The devices were fixed to the wall 1.2 m above the floor so that the samples were taken at the height of the breathing zone of a seated person. Two devices were installed in both classrooms to guarantee data collection during the test.

The device in the corridor was fixed to the wall where the entrances to the classrooms were located. It was positioned more than one meter away from both doors, far from the waiting zones, and 2.5 m high to assure its security, while possible distortions caused by users' activity were avoided. The specific location of each device can be seen in Figures 3 and 4.

The measurement devices inside the building were fixed to the wall using a plastic casing made of polylactide (PLA) in 3D printing. The device used to measure the outdoor conditions was placed on the roof of the building using a support that raised it 0.5 m from the deck. It was protected from weather conditions with a perforated casing, so air circulation was enabled.

The students, teachers, and administrative staff were only informed about the general functioning of the devices and were asked to stay away from them to avoid distortions. In this way, users' behavior was not conditioned concerning ventilation performance in order to maintain normal operational conditions.

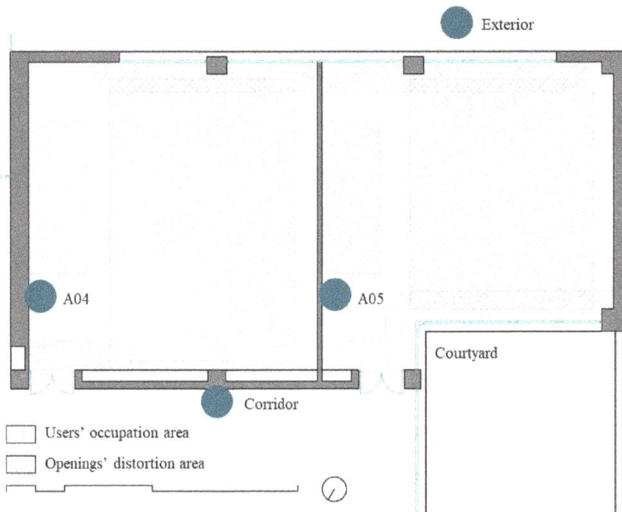

Figure 3. Locations of the measurement devices during the test.

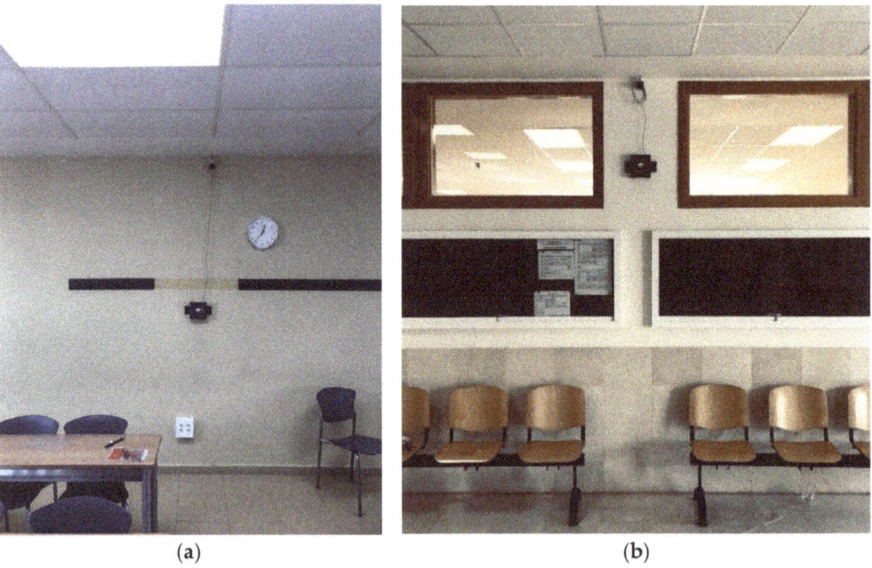

(a) (b)

Figure 4. Locations of the measurement devices in (**a**) A04, (**b**) corridor.

2.4. Airtightness and Air Infiltration

Pressurization tests were performed in classrooms A04 and A05 according to ISO 9972 [37] in order to assess the airtightness of the building envelope that allows infiltration airflows. Indications described for Method 2, which tested the airtightness of the building envelope, were followed. For this purpose, only windows and doors were closed since the classrooms had no other kind of intentional openings or air conditioning system. Semi-automatic tests were carried out generating both an overpressure and a depression in the spaces to be tested. The air change rate (n_{50}), the air permeability rate (q_{50}), and the pressure exponent n, which can be considered as an indicator of the airflow regime, were obtained from the power law equation [38]. The correct calibration of the equipment was ensured to maintain accuracy specifications of 1% of sampling, or 0.15 Pa.

To characterize both the global airtightness and interzonal leakages, three tests were performed for each classroom: individual classroom test (A04 and A05), guard-zone pressure test (simultaneously pressurizing the upper classroom, 1.04 and 1.05), and guard-zone test (simultaneously pressurizing the adjacent classroom). Airflow rates of the guard-zone tests were deducted from those of the single-unit test in order to determine the proportion of leakages between classrooms. However, it was not possible to carry out guard-zone tests in circulation areas due to technical limitations. Consequently, the rate of air leakage of the external envelope and walls in contact with circulation areas could not be discriminated.

Air infiltration under natural conditions was estimated from the airtightness results using a simplified model [39] that assumed a linear relationship between the air change rate at 50 Pa (n_{50}) and the mean annual air change rate (ACH). The model was adapted to the climate characteristics of the location under study and building parameters. Interzonal leakages were considered negligible due to similar pressure and temperature conditions and taking into account that interzonal airflows cannot be considered a contribution to ventilation.

2.5. Ventilation Efficiency

The ventilation efficiency of classroom A04 was analyzed using CFD software Ansys Fluent 19.0. The evaluation of the ventilation efficiency encompasses the age of the air of the masses that move around the space, identifying air stagnation areas within the room. The air inside a building is progressively polluted from its inlet to its exhaust as a consequence of different sources such as occupants' breathing, furniture, equipment, or other pollutants emissions.

The method used to obtain the values of the mean age of the air $\langle \overline{\tau} \rangle$ and the local age of the air in the exhaust surface τ_e was defined by [40,41]. The ventilation efficiency ε was obtained from the relationship between both values [42] (Equation (1)).

$$\varepsilon = \frac{\tau_e}{2 \cdot \langle \overline{\tau} \rangle} \qquad [\%]. \tag{1}$$

The isothermal model used for this purpose was preprocessed using the parameters described in Table 2. The obtained mesh had a skewness of over 90%.

Table 2. Mesh parameters.

Y+ (Non-Dimensional Mesh Distance to Subviscous Layer)	<30
d (distance from the wall to the first cell)	<0.01
Maximum cell proportion	1:4
Maximum growth between cell nodes	10%
Number of cells	6,116,846

Reynolds-averaged Navier Stokes (RANS) equations [43] were applied to simulate the airflow and air mass movement along the classroom. The numerical simulation was carried out in a four-stage approximation, using the k-ε equations with Standard Wall Treatment proceeding, Renormalization Group (RNG) with standard and near wall approximation, and Realizable with Enhanced Wall Treatment. Each step was accepted after convergence criteria of 10^{-4} for air velocity, momentum and turbulence parameters (k-ε). The values of the local age of the air were also obtained by applying the physics related to the characterization of the model through User Defined Functions (UDFs). Convergence criteria for the age-of-the-air value defined by the UDF was fixed under 10^{-5}. The final simulation took approximately 18,000 iterative cycles. The isothermal setting accuracy had been previously validated [44].

Boundary conditions of the model were defined according to the built model. Walls, ceiling, and floor were defined as ideally airtight boundaries, while envelope window perimeters were defined as velocity inlet and the gap under the door as pressure outlet. Velocity inlet boundary type was set up as a vectorial air velocity, including a maximum turbulence distortion length of 14 mm according to the gap.

The airflow consigned to the classroom was obtained from the annual air change rate estimated from the airtightness tests results, excluding interzonal leakages. The air inlets were located along the envelope window perimeters, according to leakage identification through thermal imaging. The openings (windows and doors) of the classrooms were assumed to be closed, and all the architectural elements (leakage paths, materials, and envelope) were assumed the most unfavorable conditions.

3. Results

3.1. IAQ Results

The average IAQ values registered in both classrooms during the occupied period are shown in Table 3.

Table 3. Average values of Indoor Air Quality (IAQ) parameters.

	Classroom	Occupation Time (h)	T_{int} (°C)	T_{ext} (°C)	RH (%)	CO_2 (ppm)	CO_2 EXT. (ppm)	CO_2 CORR. (ppm)	PM1.0 (μg/m^3)	PM2.5 (μg/m^3)	PM10.0 (μg/m^3)	TVOC (ppb)
Monday	A04	6	23.2	8.0	36.6	1561	374.5	1186	0.6	0.7	0.8	593.6
	A05	8	22.5		40.3	1817		1442	0.0	0.0	0.1	642.1
Tuesday	A04	4	23.7	8.0	40.5	1506	379.0	1127	0.0	0.1	0.1	559.4
	A05	6	22.4		41.4	1327		948	0.0	0.1	0.2	474.2
Wednesday	A04	5	23.1	11.5	44.1	1339	383.4	955	0.1	0.3	0.5	520.4
	A05	6	22.2		47.2	1800		1416	0.0	0.3	0.7	522.2
Thursday	A04	2	22.6	7.0	46.8	1330	421.0	909	0.1	0.4	0.5	447.0
	A05	6	22.4		43.5	1327		906	0.0	0.2	0.4	364.8
Friday	A04	4	24.2	17.4	34.6	1292	382.8	910	3.2	5.6	6.7	579.5
	A05	3	23.7		42.1	2313		1931	2.4	4.2	5.6	805.3
Summary	A04	21	23.4	10.4	40.5	1406	385.7	1017	0.8	1.4	1.7	540.0
	A05	29	22.6		42.9	1717		1328	0.5	1.0	1.4	561.7

A relationship between the increase in the levels of pollutants and occupation in the classrooms during teaching hours can be observed, especially concerning CO_2 and TVOC concentration. The temperature also increased slightly when the classrooms were occupied.

The average temperature was 23.4 and 22.6 °C for classrooms A04 and A05, respectively. In the case of classroom A05, the temperature was within the range set by RITE [33]. The scenario was different in classroom A04, where the average temperature was above the limit value. Temperatures were under 21 °C for less than 1% of the whole time in the case of classroom A04 and 22% of the time in classroom A05. The maximum temperature limit was exceeded for 74.5% of the time in classroom A04, and 22% in classroom A05.

In the case of RH, the average values registered during the week were 40.5% for classroom A04 and 42.9% for classroom A05. RH was under the optimal range [33] for 53% of the time when classroom A04 was occupied and 26% in the case of classroom A05. The limits were exceeded less than 1% of the time in the case of classroom A04 and 12% in classroom A05.

The CO_2 concentration outside the building was between 369 and 441 ppm, and the average concentration throughout the week was 386 ppm. In order to obtain a more accurate CO_2 correction, the CO_2 average per day was used. The average CO_2 concentration registered in the corridor was 608 ppm. Users' behavior regarding the opening of both doors and windows can be observed in the results since it caused the decrease of the concentration of the pollutants even when the classroom was occupied. It was possible to observe this reduction or stabilization even during the breaks between lessons.

The average values of CO_2 corrected concentration were 1017 ppm in the case of classroom A04 and 1328 ppm in classroom A05, although Spanish regulations set a maximum value of 500 ppm [33]. This means that the values exceeded the maximum limit for 88% of the time in classroom A04 and 74% in the case of A05. Values that doubled the maximum allowed limits were reached 62% of the time, reaching even four times above the limits for 25% of the time that the students were inside classroom A05. The highest value registered during the week was on Wednesday in classroom A05 at 11:50, only two hours after the beginning of the classes, when the CO_2 corrected concentration was 2871 ppm.

The values registered in the case of PM were not significant. The mean concentration of $PM_{1.0}$, $PM_{2.5}$, and $PM_{10.0}$ was below 7 $\mu g/m^3$ in both classrooms during the whole duration of the measurements when the classrooms were occupied. However, it should be highlighted that PM levels registered outdoors were slightly higher than indoors, likely motivated by other activities surrounding the building, such as a construction site. The peak values registered outside the building, especially on Friday, had an impact on the registered PM results inside the classrooms, which were in any case within acceptable ranges.

The average values registered in the case of TVOC concentration were 540 and 562 ppb in classrooms A04 and A05, respectively. During the teaching period, the concentration of this pollutant was over 500 ppb 64% of the time in classroom A04 and 57% in classroom A05. The highest values registered were 973 ppb in classroom A04 and 1044 ppb in classroom A05, both registered on Monday.

3.2. Airtightness and Air Infiltration Results

The airtightness results obtained for the individual classroom tests (A04 and A05), guard-zone tests with the adjacent classroom (A04-A05 and A05-A04), and guard-zone tests with the upper classroom (A04-1.04 and A05-1.05) are shown in Table 4. Blank values are results discarded due to invalid tests that did not comply with the measurement standard.

Table 4. Pressurization tests results.

	Depressurization				Pressurization				Mean Values		
	V_{50} [m³/h]	n_{50} [h⁻¹]	q_{50} [m³/m²·h]	n [-]	V_{50} [m³/h]	n_{50} [h⁻¹]	q_{50} [m³/m²·h]	n [-]	V_{50} [m³/h]	n_{50} [h⁻¹]	q_{50} [m³/m²·h]
A04	7414	24.1	22.5	0.74	4710	15.3	14.3	0.64	6062	19.7	18 4
A04-A05	2817	9.2	8.6	0.50	-	-	-	-	2817	9.2	8.6
A04-1.04	7864	25.6	23.9	0.74	9398	30.6	28.6	0.99	9631	28.1	26.2
A05	6635	24.2	23.1	0.70	8395	30.6	29.3	0.86	7515	27.4	26.2
A05-A04	2399	8.7	8.4	0.67	3352	12.2	11.7	0.93	2876	10.5	10.0
A05-1.05	6535	23.8	22.8	0.72	8346	30.4	29.1	0.96	7440	27.1	25.9

In the individual tests carried out in each classroom, the average air change rate at a reference pressure of 50 Pa (n_{50}) obtained in classroom A04 was 19.7 h⁻¹, while in classroom A05 it was considerably higher: 27.4 h⁻¹. The dispersion of the results obtained at depression and overpressure should be highlighted. Additionally, great variability of the pressure exponent n could be observed (between 0.64 and 0.86).

Results regarding the interzonal airflow rates obtained from the guard-zone pressurization tests are shown in Table 5.

Table 5. Guard-zone pressurization results.

	Depressurization				Pressurization				Mean Values			
	V_{50} [m³/h]	n_{50} [h⁻¹]	q_{50} [m³/m²·h]	%	V_{50} [m³/h]	n_{50} [h⁻¹]	q_{50} [m³/m²·h]	%	V_{50} [m³/h]	n_{50} [h⁻¹]	q_{50} [m³/m²·h]	%
A04-A05	4597	14.9	14.0	62	-	-	-	-	3245	10.5	10.5	53.5
A04-1.04	-	-	-	-	-	-	-	-	-	-	-	-
A04 total	4597	14.9	14.0	62	-	-	-	-	3245	10.5	10.5	53.5
A05-A04	4236	15.4	14.8	63.8	5043	18.4	17.6	60	4639	16.9	16.2	61.7
A05-1.05	100	0.4	0.4	1.5	49	0.2	0.2	0.6	75	0.3	0.3	1
A05 total	4336	15.8	15.1	65.3	5092	18.5	17.8	60.6	4714	17.2	16.4	62.7

The proportion of the infiltration rate at 50 Pa between classrooms A04 and A05 was between 53.5 and 63.8%. In the case of interzonal infiltration between the classrooms under study and the upper floor classrooms, the proportion seemed to be significantly reduced to an average of 1%. Therefore, the infiltration rate between adjacent classrooms represented up to 65.3% of the total air infiltration.

The air change rate under natural pressure conditions ACH [h^{-1}] due to air infiltration obtained, excluding infiltration between classrooms, was 0.44 h^{-1} in classroom A04 and 0.49 h^{-1} in classroom A05.

Thermal images taken during the depressurization stage of the airtightness test were used to locate air leakage paths of the building envelope (Figure 5). Air infiltration was identified in the perimeter of exterior windows, especially in rolling-shutter boxes and joint window–wall. In view of the numerical results obtained, air leakage paths were also expected to be found around interior windows. However, this kind of leakage could not be thermally identified given the lack of temperature gradient.

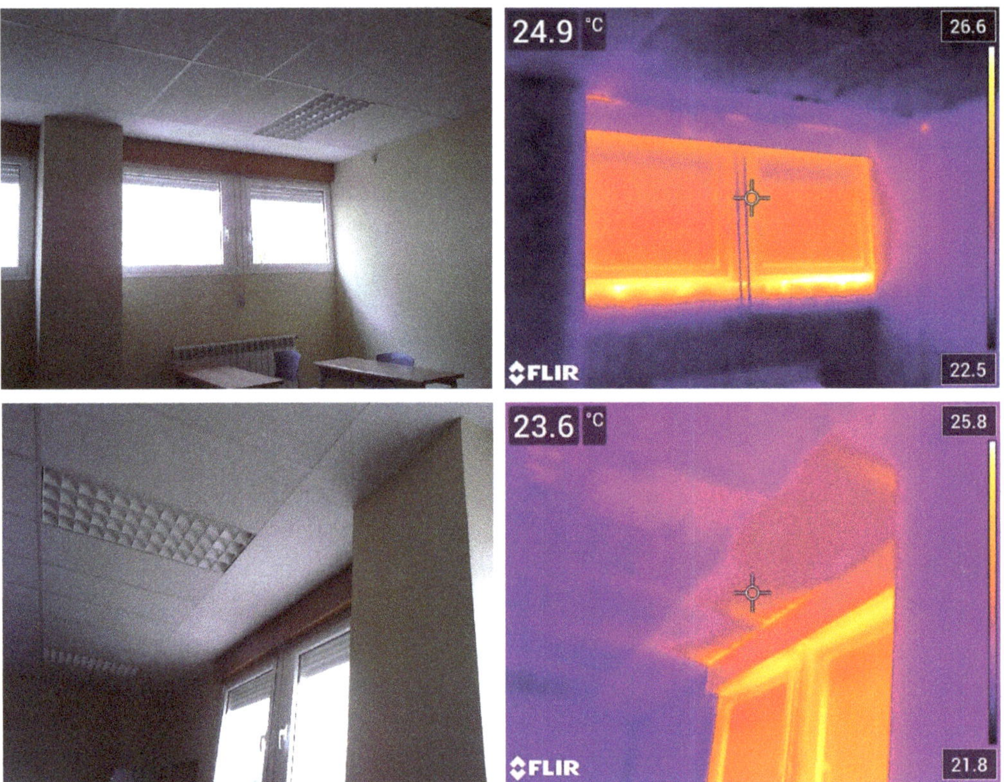

Figure 5. Air leakage paths in classroom A04 via infrared images.

3.3. Ventilation Efficiency Results

The ventilation efficiency for classroom A04 was assessed by means of CFD simulation of the isothermal model considering the estimated annual air change rate obtained (0.44 h^{-1}). This means that 44% of the air volume of the classroom (309 m^3) was changed every hour through the leakages of the envelope. Therefore, an airflow of 136 m^3/h was set up, reaching an air velocity of 2.89 m/s close to the infiltration paths (Figure 6). The model revealed that the worst scenario was detected in the wall opposite the windows, where the age of the air reached a value of 9000 s (Figure 7).

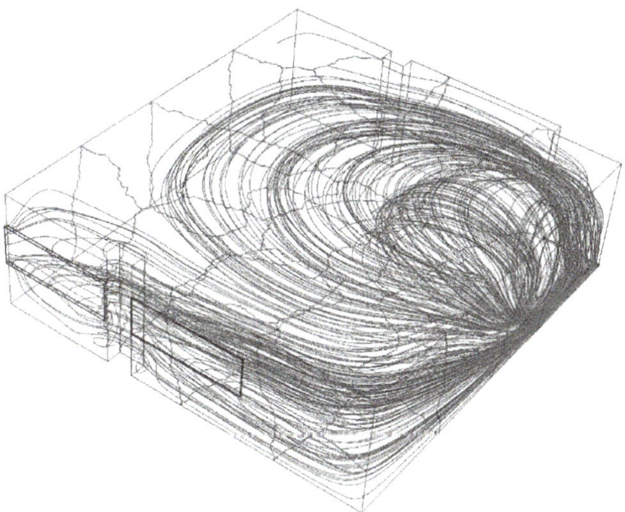

Figure 6. Air path assessment in classroom A04.

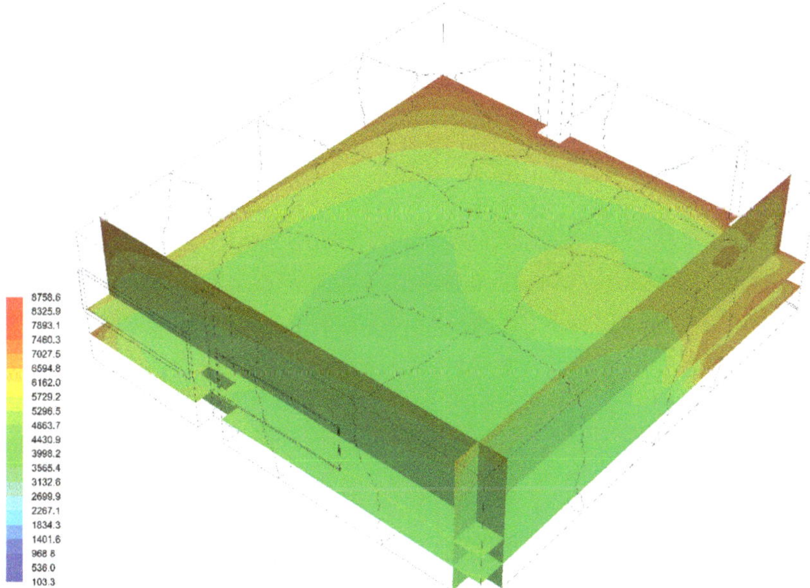

Figure 7. Age-of-the-air assessment in classroom A04.

This means that the stagnated air in this region took 2.5 h to renew, increasing the pollutants' concentration, and thus affecting the quality of the air that occupants breathe in this zone. The ventilation efficiency of the classroom obtained from the model was 53.25%. A ventilation efficiency value close to 50% represents a perfect mixing flow model, where the incoming fresh air uniformly mixes with the interior air mass, whose average residence time is twice the transit time [45,46]. The amount of 53.25% indicated that the ventilation flow rate could be reduced while maintaining the IAQ. However, the natural ventilation solution of the classroom and the poor IAQ values registered inhibited that reduction.

Figure 7 shows that most of the occupied space is reached by the ventilation airflow, so the problem regarding distribution and dilution of contaminants was considered solved.

4. Discussion

According to the high levels of pollutant concentration registered and the observation of the graphs during the teaching period, it could be deduced that the windows were hardly ever opened to ventilate the occupied classrooms in order to reduce the pollutant concentration associated with the use of some materials during academic activity and human presence.

CO_2 concentration monitoring offered valuable information as an indicator of ventilation performance. High values were reached in both classrooms even though the occupation of both spaces was kept below their maximum capacity, highlighting the poor ventilation performance. The data showed that windows were not opened during the teaching period despite the high values reached. The CO_2 concentration graphs (Figure 8) show its rapid increase after the students entered the classrooms, registering high values during the entire teaching period, which can constitute a risk for the health and academic performance of the students. It was possible to observe as well the rapid reduction of CO_2 concentration due to the opening of doors and, in some cases, windows, and when the students left the classroom. In spite of that, the decrease did not mean the recovery of acceptable values, and there was an increase in the pollutant when the classroom was occupied again.

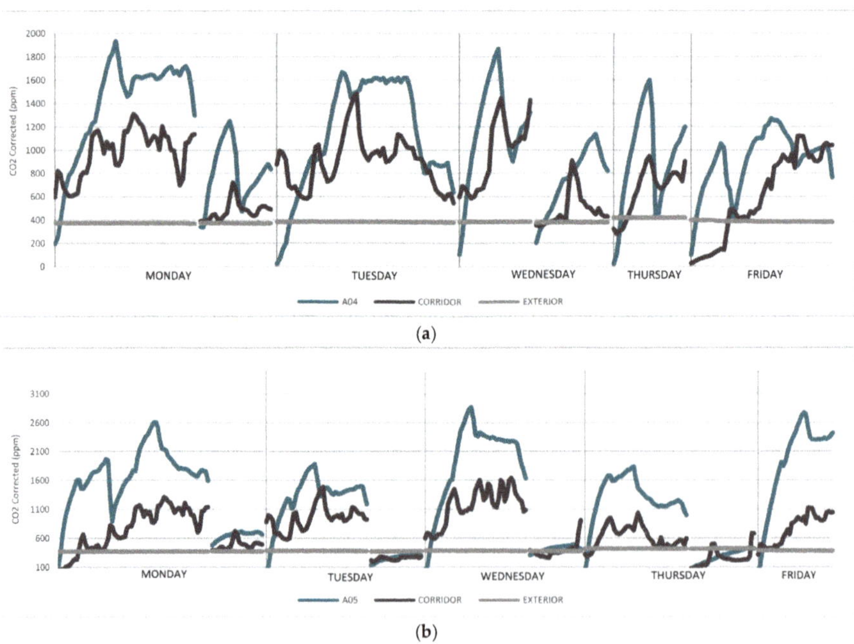

Figure 8. CO_2 measurements during the week in (a) A04 (b) A05.

At the end of the day and during the lunch break, there was a decrease of the CO_2 levels registered in the classroom, being below or equal to the outside CO_2 concentration. This means that the concentration at the beginning of each shift was adequate under the maximum accepted values and there was no accumulation of the pollutants from one day to the following one. Therefore, only poor ventilation of the spaces was responsible for these high values registered when the classrooms were occupied (Figure 8).

This was a consequence of the randomness of natural ventilation due to two factors. The first one was that natural ventilation depends on physical conditions such as interior and exterior temperature or wind direction. The second factor was that it depends as well on the users' decisions, which are not based on objective criteria. Another problem associated with natural ventilation and air infiltration was energy loss during the cold season. This was probably one of the reasons users decided not to open the windows.

In the case of TVOC concentration, trends were similar to CO_2 concentration, registering strong increases when students occupied the classrooms. However, more variations and peaks were registered. The TVOC concentration inside classes at the beginning of each day as well as at the beginning of the first class in the afternoon was equal to or lower than the outdoor concentration, and TVOC concentration in the corridor remained higher (Figure 9).

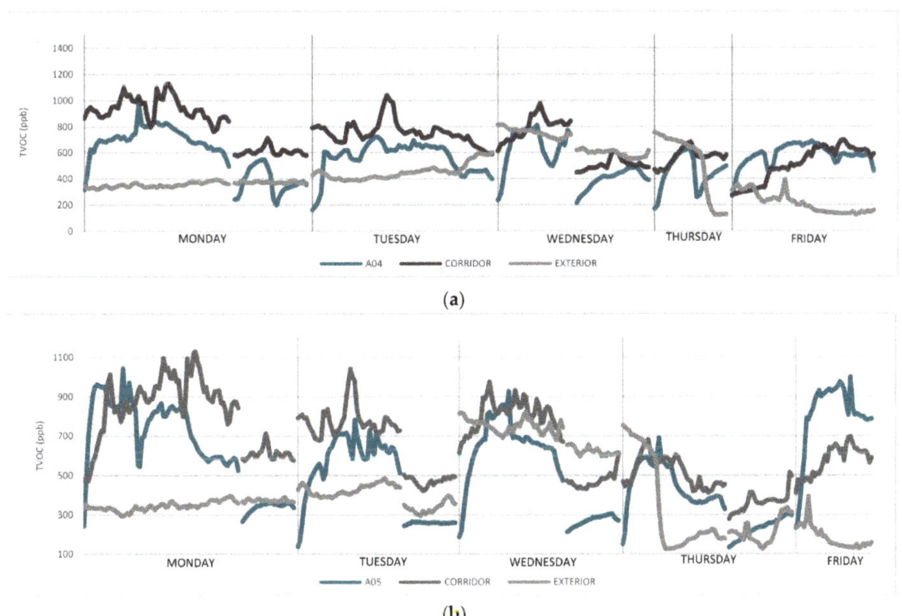

Figure 9. Total Volatile Organic Compounds (TVOC) concentration measurements during the week (**a**) A04 (**b**) A05.

Pressurization tests performed in both classrooms revealed that the airtightness of the building envelope in both classrooms was rather poor, with values of the air change rate n_{50} greater than 20 h^{-1} (except for the depressurization mode in classroom A04). The interzonal leakage between classrooms on the same floor was very high, constituting an average of 58% of the total airflow at 50 Pa, while leakages between floors could be considered negligible. Although this significant permeability between classrooms did not impact the energy demand, it must be considered in terms of pollutant transmission, comfort, and noise.

The infiltration air change rate (ACH) of the classrooms, around 0.47 h^{-1}, contributed to the renewal of the air in the classrooms. However, despite the poor airtightness of the envelope, air infiltration is an uncontrolled phenomenon and hence cannot be relayed as a single ventilation source confirmed by poor IAQ results. According to ventilation requirements established in the Spanish regulations for educational buildings based on theoretical occupation, air infiltration should contribute less than 5% of the total required ventilation rate [33]. Nevertheless, the ventilation efficiency obtained guaranteed a good homogeneity of the airflow along the classroom, assuming a perfect mix of the air ($\varepsilon \approx 50\%$) due to the inlets and exhaust location.

5. Conclusions

The IAQ of two classrooms with a natural ventilation system was assessed during the cold season. PM levels were always within acceptable ranges, while CO_2 and TVOC concentration exceeded recommended concentration levels. Their increase was related to the occupation of the classrooms by the students.

The CO_2 concentration was within the accepted range only 26% of the time, while TVOC concentration was only acceptable 43% of the time in the most favorable scenario. These results are worrying, especially if it is considered that both classrooms were below their maximum capacity. The opening of doors and windows favors a decrease of the concentration of the pollutants, but users rarely opened them during the teaching period. Even though the acceptable limits recovered every day after the users left the building, it only took a few minutes to exceed them when the classes started the following day.

Therefore, it is undeniable that the air renewal contributed by natural ventilation and air infiltration was not enough to maintain good IAQ. However, the obtained ventilation efficiency of one of the analyzed classrooms (53.25%) was adequate for its current configuration. The ventilation performance was poor due to the randomness of both air infiltration and natural ventilation, which also encompass energy loss during the cold season, whose estimation was not the object of this study.

All things considered, the improvement of the ventilation performance of the classrooms and the airtightness of the envelope should be pursued, maintaining at the same time temperature and RH comfort conditions, to guarantee the health and good academic performance of the students. Implementing a specific controlled ventilation system adapted to the case would improve IAQ and contribute to reducing the energy demand as a consequence of a minoration of the airflow needed.

The best scenario assumes an equilibrated model, in which the inlet air and exhaust air have the same flow. Inlet and exhaust positions should promote cross ventilation, reducing the stagnation and contributing to avoid air infiltration. Nevertheless, it must be acknowledged that any change in the conditions would alter the air pattern affecting the ventilation efficiency, so a further evaluation of the ventilation system would be needed.

Author Contributions: Conceptualization, I.P.-C. and A.M.; methodology, M.Á.P.-M.; formal analysis, I.P.-C.; investigation, R.G.-V.; resources, R.G.-V.; data curation, A.M.; writing—original draft preparation, R.G.-V. and I.P.-C.; writing—review and editing, M.Á.P.-M.; visualization, A.M.; supervision, M.Á.P.-M.; project administration, M.Á.P.-M.; funding acquisition, M.Á.P.-M. All authors have read and agreed to the published version of the manuscript.

Funding: This research was funded by the University of Valladolid in collaboration with ARCOR, S.L., and Hermanos Rubio Grupo Constructor HERCE, S.L.U., grant number 18IQBC. The APC was funded by the same team.

Acknowledgments: This work was supported by the University of Valladolid under the research project "REVEDUVa: Energy recovery through ventilation of university classrooms", framed within the R&D projects on energy efficiency measures and the application of renewable energies in the operation of the university buildings of the University of Valladolid in collaboration with ARCOR, S.L., and Hermanos Rubio Grupo Constructor HERCE, S.L.U. The authors would also like to thank the University of Valladolid for the funding of the doctoral program of one of the authors.

Conflicts of Interest: The authors declare no conflict of interest.

References

1. Awbi, H.B. *Ventilation of Buildings*; Spon Press: London, UK, 2004. [CrossRef]
2. Bakke, J.V.; Norbäck, D.; Wieslander, G.; Hollund, B.E.; Florvaag, E.; Haugen, E.N.; Moen, B.E. Symptoms, complaints, ocular and nasal physiological signs in university staff in relation to indoor environment—Temperature and gender interactions. *Indoor Air* **2008**, *18*, 131–143. [CrossRef] [PubMed]

3. Tham, K.W. Indoor air quality and its effects on humans—A review of challenges and developments in the last 30 years. *Energy Build.* **2016**, *130*, 637–650. [CrossRef]
4. Mujan, I.; Anđelković, A.S.; Munćan, V.; Kljajić, M.; Ružić, D. Influence of indoor environmental quality on human health and productivity—A review. *J. Clean. Prod.* **2019**, *217*, 646–657. [CrossRef]
5. Wargocki, P.; Porras-Salazar, J.A.; Contreras-Espinoza, S.; Bahnfleth, W. The relationships between classroom air quality and children's performance in school. *Build. Environ.* **2020**, *173*, 106749. [CrossRef]
6. CEN (European Committee for Standardization). *Ventilation for Non-Residential Buildings—Performance Requirements for Ventilation and Room Conditioning Systems*; CEN: Brussels, Belgium, 2008; pp. 1–75.
7. Ministerio de Industria Energía y Turismo del Gobierno de España. Real Decreto 1826/2009, de 27 de Noviembre, por el que se modifica el Reglamento de Instalaciones Térmicas en Los Edificios, Aprobado Por Real Decreto 1027/2007, de 20 de Julio. 2009. Available online: https://www.boe.es/eli/es/rd/2009/11/27/1826 (accessed on 17 June 2021).
8. Ministerio de Transportes Movilidad y Agenda Urbana del Gobierno de España, ERESEE 2020. Actualización 2020 de la Estrategia a largo Plazo para la Rehabilitación Energética en el Sector de la Edificación en España. 2020. Available online: https://www.mitma.gob.es/el-ministerio/planes-estrategicos/estrategia-a-largo-plazo-para-la-rehabilitacion-energetica-en-el-sector-de-la-edificacion-en-espana (accessed on 17 June 2021).
9. Fernández-Agüera, J.; Campano, M.Á.; Domínguez-Amarillo, S.; Acosta, I.; Sendra, J.J. CO2 Concentration and Occupants' Symptoms in Naturally Ventilated Schools in Mediterranean Climate. *Buildings* **2019**, *9*, 197. [CrossRef]
10. Krawczyk, D.A.; Rodero, A.; Gładyszewska-Fiedoruk, K.; Gajewski, A. CO2 concentration in naturally ventilated classrooms located in different climates—Measurements and simulations. *Energy Build.* **2016**, *129*, 491–498. [CrossRef]
11. Turner, W.J.N.; Sherman, M.H.; Walker, I.S. Infiltration as Ventilation: Weather-Induced Dilution. *HVAC&R Res.* **2012**, *18*, 1122–1135. [CrossRef]
12. Gomes, J.F.; Bordado, J.C.M.; Sarmento, G.; Dias, J. Measurements of Indoor Air Pollutant Levels in a University Office Building. *J. Green Build.* **2007**, *2*, 123–129. [CrossRef]
13. Burdova, E.K.; Vilcekova, S.; Meciarova, L. Investigation of Particulate Matters of the University Classroom in Slovakia. *Energy Procedia* **2016**, *96*, 620–627. [CrossRef]
14. di Giulio, M.; Grande, R.; di Campli, E.; di Bartolomeo, S.; Cellini, L. Indoor air quality in university environments. *Environ. Monit. Assess.* **2010**, *170*, 509–517. [CrossRef]
15. Righi, E.; Aggazzotti, G.; Fantuzzi, G.; Ciccarese, V.; Predieri, G. Air quality and well-being perception in subjects attending university libraries in Modena (Italy). *Sci. Total Environ.* **2002**, *286*, 41–50. [CrossRef]
16. Astolfi, A.; Corgnati, S.P.; Verso, V.R.M.L. Environmental comfort in university classrooms—thermal, acoustic, visual and air quality aspects. *Res. Build. Phys.* **2020**, 945–957. [CrossRef]
17. Lee, M.C.; Mui, K.W.; Wong, L.T.; Chan, W.Y.; Lee, E.W.M.; Cheung, C.T. Student learning performance and indoor environmental quality (IEQ) in air-conditioned university teaching rooms. *Build. Environ.* **2012**, *49*, 238–244. [CrossRef]
18. Sarbu, I.; Pacurar, C. Experimental and numerical research to assess indoor environment quality and schoolwork performance in university classrooms. *Build. Environ.* **2015**, *93*, 141–154. [CrossRef]
19. Satish, U.; Mendell, M.J.; Shekhar, K.; Hotchi, T.; Sullivan, D.; Streufert, S.; Fisk, W.J. Is CO2 an Indoor Pollutant? Direct Effects of Low-to-Moderate CO 2 Concentrations on Human Decision-Making Performance. *Environ. Health Perspect.* **2012**, *120*, 1671–1677. [CrossRef]
20. Ramalho, O.; Wyart, G.; Mandin, C.; Blondeau, P.; Cabanes, P.-A.; Leclerc, N.; Mullot, J.-U.; Boulanger, G.; Redaelli, M. Association of carbon dioxide with indoor air pollutants and exceedance of health guideline values. *Build. Environ.* **2015**, *93*, 115–124. [CrossRef]
21. Asif, A.; Zeeshan, M. Indoor temperature, relative humidity and CO_2 monitoring and air exchange rates simulation utilizing system dynamics tools for naturally ventilated classrooms. *Build. Environ.* **2020**, *180*, 106980. [CrossRef]
22. Becerra, J.A.; Lizana, J.; Gil, M.; Barrios-Padura, A.; Blondeau, P.; Chacartegui, R. Identification of potential indoor air pollutants in schools. *J. Clean. Prod.* **2020**, *242*, 118420. [CrossRef]
23. Dorizas, P.V.; Assimakopoulos, M.-N.; Helmis, C.; Santamouris, M. An integrated evaluation study of the ventilation rate, the exposure and the indoor air quality in naturally ventilated classrooms in the Mediterranean region during spring. *Sci. Total Environ.* **2015**, *502*, 557–570. [CrossRef] [PubMed]
24. Perez, P.P. *Documentos Técnicos de Instalaciones en la Edificación. DITE 2.02. Calidad de Aire Interior*; ATECYR: Madrid, Spain, 2006.
25. Daisey, J.M.; Angell, W.J.; Apte, M.G. Indoor air quality, ventilation and health symptoms in schools: An analysis of existing information. *Indoor Air* **2003**, *13*, 53–64. [CrossRef] [PubMed]
26. Almeida, R.M.S.F.; Pinto, M.; Pinho, P.G.; de Lemos, L.T. Natural ventilation and indoor air quality in educational buildings: Experimental assessment and improvement strategies. *Energy Effic.* **2017**, *10*, 839–854. [CrossRef]
27. Heracleous, C.; Michael, A. Experimental assessment of the impact of natural ventilation on indoor air quality and thermal comfort conditions of educational buildings in the Eastern Mediterranean region during the heating period. *J. Build. Eng.* **2019**, *26*, 100917. [CrossRef]
28. Lee, K.S.; Choi, S.H. Effect of geometric parameters on ventilation performance in a dry room. *Dry. Technol.* **2002**, *20*, 1445–1461. [CrossRef]

29. Li, X.; Wang, X.; Li, X.; Li, Y. Investigation on the relationship between flow pattern and air age. In Proceedings of the Sixth International IBPSA Conference-Building Simulation, Kyoto, Japan, 13–15 September 1999; Volume 99, pp. 423–429.

30. Met Office, National Meteorological Library and Archive Fact sheet 6—The Beaufort Scale. 2010. Available online: https://www.metoffice.gov.uk/binaries/content/assets/metofficegovuk/pdf/research/library-and-archive/library/publications/factsheets/factsheet_6-the-beaufort-scale.pdf (accessed on 17 June 2021).

31. Farrás, J.G. Control ambiental en interiores. In *Encicl. La Salud y Segur. en el Trab*; Instituto Nacional de Seguridad e Higiene en el Trabajo (INSHT): Madrid, España, 2012.

32. American Society of Heating Refrigerating and Air-Conditioning Engineers. *ASHRAE Handbook—Fundamentals*; American Society of Heating, Refrigerating and Air-Conditioning Engineers: Atlanta, GA, USA, 2009.

33. Ministerio de Industria Energía y Turismo del Gobierno de España. Reglamento de Instalaciones Térmicas en los Edificios. 2013. Available online: https://www.boe.es/diario_boe/txt.php?id=BOE-A-2013-3905 (accessed on 17 June 2021).

34. Umweltbundesamtes, B.; der Ad-hoc-arbeitsgruppe, H.; Innenraumlufthygiene-kommission, D. Beurteilung von Innenraum-luftkontaminationen mittels Referenz- und Richtwerten. *Bundesgesundheitsblatt Gesundheitsforsch. Gesundheitsschutz.* **2007**, *50*, 990–1005. [CrossRef] [PubMed]

35. Abdul–Wahab, S.A.A.; En, S.C.F.; Elkamel, A.; Ahmadi, L.; Yetilmezsoy, K. A review of standards and guidelines set by international bodies for the parameters of indoor air quality. *Atmos. Pollut. Res.* **2015**, *6*, 751–767. [CrossRef]

36. World Health Organization. *Air Quality Guidelines*; Global Update 2005; World Health Organization: Copenhagen, Denmark, 2006. [CrossRef]

37. International Organization for Standardization, ISO 9972:2015 Thermal Performance of Buildings. Determination of Air Permeability of Buildings. Fan Pressurization Method. 2015. Available online: https://www.iso.org/standard/55718.html (accessed on 17 June 2021).

38. American Society of Heating Refrigerating and Air-Conditioning Engineers. ASHRAE Handbook—Fundamentals. 2001. Available online: https://www.ashrae.org/technical-resources/ashrae-handbook/description-2017-ashrae-handbook-fundamentals (accessed on 17 June 2021).

39. Sherman, M.H. Estimation of infiltration from leakage and climate indicators. *Energy Build.* **1987**, *10*, 81–86. [CrossRef]

40. Meiss, A.; Feijó, J. Influencia de la ubicación de las aberturas en la eficiencia de la ventilación en viviendas. *Inf. Construcción* **2011**, *63*, 53–60. [CrossRef]

41. Sandberg, M.; Sjöberg, M. The use of moments for assessing air quality in ventilated rooms. *Build. Environ.* **1983**, *18*, 181–197. [CrossRef]

42. Etheridge, D.; Sandberg, M. *Building Ventilation: Theory and Measurement*; Wiley: Chichester, UK, 1996.

43. Yakhot, V.; Orszag, S.A. Renormalization group analysis of turbulence. I. Basic theory. *J. Sci. Comput.* **1986**, *1*, 3–51. [CrossRef]

44. Meiss, A.; Feijó-Muñoz, J.; García-Fuentes, M.A. Age-of-the-air in rooms according to the environmental condition of temperature: A case study. *Energy Build.* **2013**, *67*, 88–96. [CrossRef]

45. Liddament, M.W. *Technical Note AIVC 21. A Review and Bibliography of Ventilation Effectiveness-Detinitions, Measurements, Design and Calculation*; International Energy Agency: Bracknell, UK, 1987.

46. Skaret, E. Ventilation by Displacement—Characterization and Design Implications, Vent. '85 (Chemical Eng. Monogr. 24). 1986, pp. 827–841. Available online: https://www.aivc.org/resource/ventilation-displacement-characterization-and-design-implications (accessed on 17 June 2021).

 *sustainability*

MDPI

Article

Fique as a Sustainable Material and Thermal Insulation for Buildings: Study of Its Decomposition and Thermal Conductivity

Gabriel Fernando García Sánchez [1], Rolando Enrique Guzmán López [2,*] and Roberto Alonso Gonzalez-Lezcano [3]

[1] Grupo de Investigación en Energía y Medio Ambiente (GIEMA), School of Mechanical Engineering, Universidad Industrial de Santander—UIS, Bucaramanga 680002, Colombia; g.garciasanchez@yahoo.es
[2] Grupo de Investigación en Desarrollo Tecnológico, Mecatrónica Y Agroindustria (GIDETECHMA), Universidad Pontificia Bolivariana—UPB, Bucaramanga 680002, Colombia
[3] Architecture and Design Department, Escuela Politécnica Superior, Campus Montepríncipe, Universidad San Pablo CEU, CEU Universities, Boadilla del Monte, 28668 Madrid, Spain; rgonzalezcano@ceu.es
* Correspondence: rolando.guzman@upb.edu.co

Abstract: Buildings consume a large amount of energy during all stages of their life cycle. One of the most efficient ways to reduce their consumption is to use thermal insulation materials; however, these generally have negative effects on the environment and human health. Bio-insulations are presented as a good alternative solution to this problem, thus motivating the study of the properties of natural or recycled materials that could reduce energy consumption in buildings. Fique is a very important crop in Colombia. In order to contribute to our knowledge of the properties of its fibers as a thermal insulator, the measurement of its thermal conductivity is reported herein, employing equipment designed according to the ASTM C 177 standard and a kinetic study of its thermal decomposition from thermogravimetric data through the Coats–Redfern model-fitting method.

Keywords: thermal insulation; sustainable materials; fique; thermal conductivity; thermogravimetry; green architecture; thermogravimetry

Citation: García Sánchez, G.F.; Guzmán López, R.E.; Gonzalez-Lezcano, R.A. Fique as a Sustainable Material and Thermal Insulation for Buildings: Study of Its Decomposition and Thermal Conductivity. *Sustainability* **2021**, *13*, 7484. https://doi.org/10.3390/su13137484

Academic Editor: Antonio Caggiano

Received: 11 June 2021
Accepted: 2 July 2021
Published: 5 July 2021

Publisher's Note: MDPI stays neutral with regard to jurisdictional claims in published maps and institutional affiliations.

1. Introduction

Global warming has become one of the greatest challenges facing humans in modern times, causing serious problems such as heat waves, drinking water shortages, and the spread of disease [1]. This phenomenon is mostly caused by the increase in the concentration of greenhouse gases in the atmosphere due to the use of fossil fuels to satisfy the growing demand for energy, which is, in turn, driven by population growth, the increase in the number of vehicles, and the development of new information technologies, among other reasons. According to the Intergovernmental Panel on Climate Change (IPCC), in each of the last three decades, the Earth's surface has been successively warmer than any previous decade since 1850 [2]. To alleviate this problem, it is necessary to develop new energy sources and to find ways to reduce society's energy consumption. The latter is a major challenge for the building sector, which, together with industry and transport, is one of the most energy-intensive sectors in the world, and is increasing due to the growing use of ventilation, air conditioning, and heating systems [3]. It is estimated that buildings consume about 40% of the world's energy, 25% of the world's water, and 40% of the world's resources, and are responsible for 1/3 of the world's greenhouse gas emissions [4–6]. For this reason, there has been a great deal of interest in improving their energy efficiency around the world, which can be seen in the increase in research on the subject [7] and in the emergence of regulations [8,9].

The use of thermal insulation is recognized as one of the most efficient ways to reduce energy consumption in buildings [3,9–11]. For this purpose, materials obtained from petrochemical products (mainly polystyrene) or from processed natural sources with high

17

energy consumption (such as rock wool and glass fiber) are often used. This has negative effects on the environment, both in terms of energy consumption and the generation of waste and emissions, mainly in the production stage [3,12], as well as causing serious problems for human health [13]. In fact, some of the materials used as insulation have recently been banned because they may pose health risks [14,15]. These problems could be addressed by using insulation made from recycled or natural materials (biomass) that do not require such a high degree of processing. Unfortunately, these materials are still in the development stage and, in many cases, their most important properties have not yet been fully determined [12]. The study of biomass as an insulating material (natural insulators) began in 1974 and remained a topic of little interest, with few scientific articles published until 1998, when the number of studies published on the subject began to increase. This trend progressed even more after 2003, and especially between 2010 and the present day. Currently, there is a great deal of interest in the development of this type of thermal insulator [3,16–18]. This interest is due to the increased environmental awareness of society, the increase in air conditioning systems—which require methods to reduce energy consumption—and changes in the use of biomass, for which other alternatives are required for its efficient use, as it is being replaced by hydrocarbon fuels in applications such as heating [3]. Studies on the subject have mainly been carried out in France, the UK, Italy, Turkey, and Algeria, which are the countries with neither the largest arable land area nor the largest number of forests; this can be explained by the greater environmental awareness of European countries compared with the rest of the world. Additionally, as expected, the most researched biomasses are from crops grown in these countries, i.e., hemp, straw, flax, wood, coconut, maize, and sunflower [3]. A comprehensive review of the state-of-the-art work on the main natural insulators around the world is presented in the works of Liu et al. [3], As-drubali et al. [12], Hurtado et al. [19], Mangesh et al. [20], Ingrao et al. [21], and Kymäläinen and Sjöberg [22].

The feasibility of using fique as a thermal insulating material is a subject that has rarely been studied in the world, which is evidenced by the low number of scientific articles published on the subject. However, there has been growing interest in the subject since 2003, which has been accentuated since 2010. Some of the most relevant works on the subject are presented in Table 1. The superficial characteristics of fique fibers were studied by Guzmán et al. [23]. An expansion in the field of research towards the manufacture of composite materials (biopolymer/fiber, aerogel/fiber) is shown in the works of An et al. [24] and Dou et al. [25]. Fique is a crop of great importance for Colombia, which is the world's main producer, and the livelihoods of more than 70,000 Colombian families depend on its production [26]. In the country, the plant's fibers are widely available throughout the year [27], making it ideal for uses beyond the manufacture of ropes and garments in order to contribute to our knowledge of the properties of fique as a thermal insulator and provide a more holistic overview of the environmental impact during the whole specific life cycle phase [28,29], from raw material extraction to end of life stages; hence enabling justified decisions on the suitability of using fique fibers in thermal insulation [30]. This paper reports the measurement of its thermal conductivity, by means of the guarded hot plate method (ASTM C 177), and a study of its thermal decomposition by means of the analysis of thermogravimetric data using the Coats–Redfern method.

Table 1. Studies on the characterization of fique as a thermal insulator.

Authors	Description	Year
Muñoz y Cifuentes [31]	Study of the thermal life of containers insulated with fique and the thermal conductivity of four commercial presentations of the material (chopped, unchopped, in wadding fabric and chopped in a thermosetting resin matrix).	2007
Onésippe et al. [32]	Analysis of the mechanical and thermal behavior of a composite material made of cement reinforced with bagasse and fique fibers.	2008
Monsalve et al. [33]	Study of the mechanical and thermal behavior of a material made up of three layers: two outer layers of composite with cementitious matrix with fique fiber and aluminum oxide powder and an inner layer of recycled newsprint pulp (cellulose).	2013
Navacerrada et al. [34]	Characterization of samples made of woven and nonwoven fique of different densities and thicknesses, for which sound absorption, air flow resistivity, and thermal conductivity as a function of density were measured.	2013
Navacerrada et al. [35]	Characterization of samples made up of nonwoven fique samples with a polymeric surface coating.	2014
Proaño [36]	Development of two types of rigid polyurethane matrix composite materials, one reinforced with cabuya (fique) fibers and the other reinforced with African palm rachis fibers. Tests were carried out on bending, traction, density, combustion speed, and acoustic and thermal properties.	2015
Navacerrada et al. [37]	Determination of the acoustic and thermal properties of samples made from coconut, coconut/fique, and fique nonwoven fibers, which were produced by two methods: pressing with binder and punching.	2016
Vera [38]	Study of the acoustic and thermal performance of cabuya fiber as a wall cladding panel.	2018
García et al. [39]	Morphological study and description of the thermal decomposition process of three types of fique fiber samples: untreated, washed with a commercial softener, and after soaking for 24 h in the same softener.	2019
Gómez et al. [40]	Morphological and thermo-acoustic characterization of nonwoven fique samples. The sound absorption, resistivity, dynamic stiffness and thermal conductivity of the material were measured.	2020

2. Methodology

2.1. Measurement of Thermal Conductivity

Thermal conductivity is one of the main characteristics of thermal insulators. According to standards such as TS 825, ISO 9164:1989 and DIN 4108, materials with thermal conductivities of less than 0.07 W/m·K can be considered thermal insulators [41]; other studies consider materials to be thermal insulators if the conductivity is less than 0.1 W/m·K [14]. One of the most widely used methods to measure this property is the protected hot plate method, which consists of taking the sample to a stationary state with two different and known temperatures so that Fourier's law of heat conduction in its one-dimensional form is applicable; this can be expressed as follows:

$$k = \frac{\dot{Q}L}{A\Delta T} \tag{1}$$

where k is the thermal conductivity of the material, $\dot{Q}$ is the heat transfer rate, L is the thickness of the sample, A is the heat transfer area, and ΔT is the temperature difference. For this purpose, the sample is located between two plates that act as a heat source and a heat sink, in order that, from their temperatures, the power required by the device, its geometry, and the material's ability to conduct heat can be calculated.

In the present study, the protected hot plate method was used to measure the conductivity of nonwoven fique samples at different densities, for which the equipment shown in Figure 1 was built, which was designed according to ASTM C 177-13 [42] with a double-sided configuration to ensure a stable and constant transmission of energy by conduction perpendicular to the surfaces of the samples. This device consists of four main parts, namely: a measurement and control system, a support system, a heating system, and a cooling system, which are presented in Figure 1.

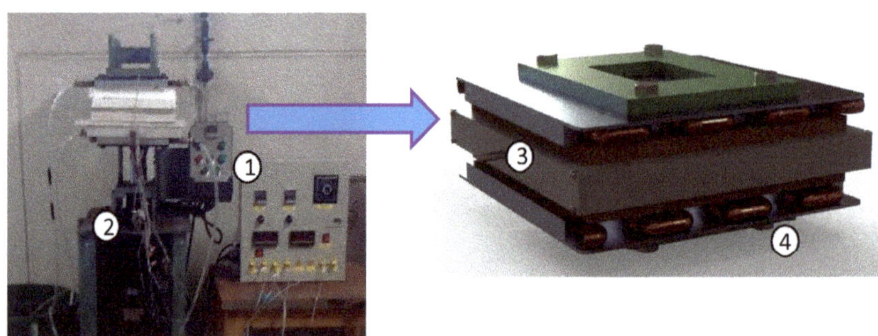

Figure 1. Thermal conductivity measuring bench in accordance with ASTM C 177-13. (1) Measuring and control system, (2) support system, (3) heating system, and (4) cooling system.

2.1.1. Measurement and Control System

The measurement and control system allows real-time temperature measurement by means of eight K-type thermocouples, four located on the hot plate (Figure 2b) and four on the cold plates (two on the upper plate and two on the inner plate). The system also supplies the electrical power required by the two flat plate heaters, which are controlled by two electronic PID microcontrollers (Autonics TCN4).

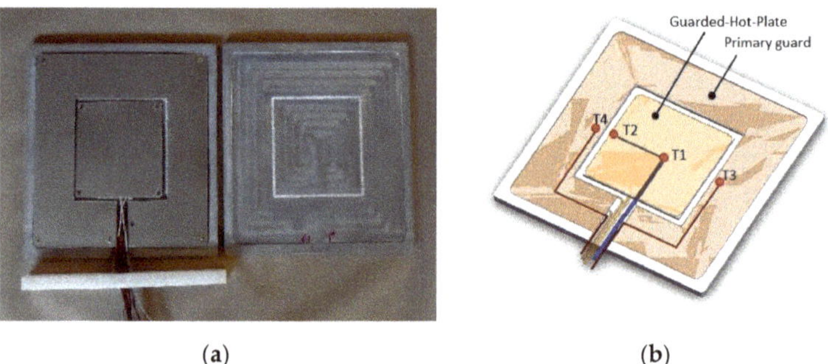

<table>
<tr><td>(a)</td><td>(b)</td></tr>
</table>

Figure 2. Heating system: (**a**) aluminum plates and electrical resistors and (**b**) thermocouples for temperature measurement in the system.

2.1.2. Support System

The support system supports the other systems and is mounted on a hydraulic unit, which ensures constant pressure on the specimens by means of a hydraulic actuator, thus ensuring the reproducibility of the measurements.

2.1.3. Heating System (Guarded-Hot-Plate)

The heating system consists of two aluminum plates, measuring 300 mm × 300 mm × 10 mm, in the middle of which there are two independent flat coplanar resistors, as shown in Figure 2; one corresponding to the area of the protected hot plate with a power value of 225 W (150 mm × 150 mm) and another resistor corresponding to the area of the primary guard, which supplies an electrical power of 360 W. Four temperature sensors (T1, T2, T3,

T4) are located on the plate to measure the temperature values and to ensure that there is no lateral flow, i.e., that there is only unidirectional flow.

2.1.4. Cooling System

The cooling system consists of two aluminum plates (cold surfaces), measuring 300 mm × 300 mm × 20 mm, through which seven $\frac{1}{2}$-inch copper tubes pass on each plate, as shown in Figure 3. Two K-type thermocouples (T5–T8) are placed on each cold plate to obtain temperature data during the test.

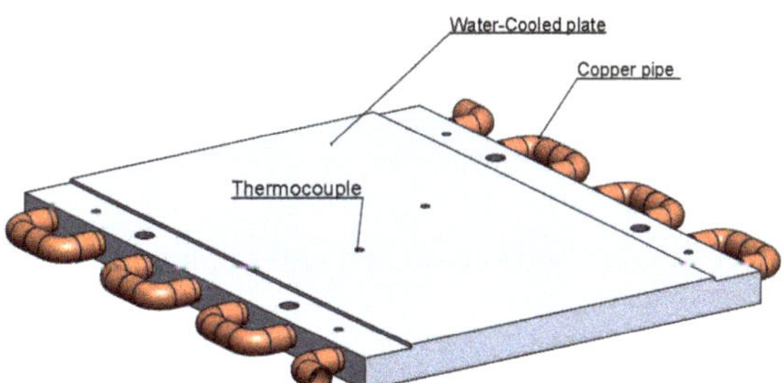

Figure 3. Cooling system assembly with two heat exchangers.

2.2. Kinetic Modelling

Kinetics, represented by the kinetic triplet, is an important property for the study of the thermal decomposition of biomass, as it allows the complete simulation of the conversion vs. time curve, as well as the control and optimization of the process parameters. The most common method to determine it is the analysis of thermogravimetric (TG) data [43,44], as it is the most effective, least expensive, and simplest way to observe fuel combustion and pyrolysis profiles [45]. The thermogravimetric (TG) tests were carried out in a TGA Q500-TA instruments analyzer, using a nitrogen atmosphere with a constant flow rate of 15 mL/min and a heating rate of 5 °C/min, taking the sample from 27 °C to 480 °C.

The rate at which the thermal decomposition process of solid samples occurs is usually expressed by Equation (2):

$$\frac{d\alpha}{dt} = f(\alpha)\, A \exp\left(-\frac{E_a}{R_u T}\right) \tag{2}$$

where $f(\alpha)$ is known as the decomposition function or reaction model, A is the pre-exponential factor, E_a the activation energy, Ru the universal pass constant, and T the absolute temperature. This equation has no analytical solution, so several methods have been developed to determine the factors $f(\alpha)$, E_a, and A (known as the kinetic triplet), which can be classified into fitting and isoconversional models. In the former, a predefined form of $f(\alpha)$ is assumed, while, in the latter, data with the same value as the degree of conversion α are selected, so that $f(\alpha)$ is constant, and A and E_a are independent of its form, allowing the Arrhenius equation to be evaluated without choosing the order of the reaction. Fitting methods have been widely used in solid-state reactions due to their ability to directly determine the kinetic parameters from TG data at a single heating rate, one of the most important of which is the Coats–Redfern integral method [46], which has been successfully used in the kinetic modelling of plant biomass, as can be seen in the work of Alvarez et al. [47] According to this method, Equation (2) has the following solution:

$$\ln\left(\frac{g(\alpha)}{T^2}\right) = \ln\left[\frac{AR_u}{\beta E_a}\left(1 - \frac{2R_u T}{E_a}\right)\right] - \frac{E_a}{R_u T} \tag{3}$$

where $g(\alpha)$ is the integral form of the reaction model and β is the reaction rate. The $2RuT/Ea$ term is very small, so it is usually neglected, leading to:

$$\ln\left(\frac{g(\alpha)}{T^2}\right) = \ln\left[\frac{AR_u}{\beta E_a}\right] - \frac{E_a}{R_u T}$$

(4)

Thus, the graph $\ln\left(\frac{g(\alpha)}{T^2}\right)$ vs. $\frac{1}{T}$ is a straight line with slope $-Ea/Ru$ and ordinate at the origin equal to $\ln\left[\frac{AR_u}{\beta E_a}\right]$, which allows us to clear the values of Ea and A. To determine the reaction model, the "master plots" graphs were used, which is the method recommended by the International Confederation for Thermal Analysis and Calorimetry (ICTAC) [48,49]. For this purpose, the reaction models presented in Table 2 were plotted first, and then the experimental curve $g(\alpha)/g(0.5)$ vs. α was plotted and the theoretical curve that best fit according to equality was chosen:

$$\frac{g(\alpha)}{g(0.5)} = \frac{\frac{E_a A}{\beta R} p(x)}{\frac{E_a A}{\beta R} p(x_{0.5})} = \frac{p(x)}{p(x_{0.5})}$$

(5)

where $x = Ea/RT$, $p(x)$ is calculated according to the approximation of Pérez-Maqueda and Criado [50]:

$$p(x) = \left(\frac{e^{-x}}{x}\right)\left(\frac{x^7 + 70x^6 + 1886x^5 + 24{,}920x^4 + 170{,}136x^3 + 577{,}584x^2 + 844{,}560x + 35{,}120}{x^8 + 72x^7 + 2024x^6 + 28{,}560x^5 + 216{,}720x^4 + 880{,}320x^3 + 1{,}794{,}240x^2 + 1{,}572{,}480x + 403{,}200}\right)$$

(6)

The experimental conversion data were compared with the respective theoretical ones obtained from kinetic modelling, and the quality of their fit was evaluated by means of the average percentage deviation (AVP) proposed by Orfao et al. [51]:

$$AVP = 100\sqrt{\frac{SS}{N}}$$

(7)

$$SS = \sum_{i=0}^{N}\left[(\alpha)_{i,exp} - (\alpha)_{i,teo}\right]^2$$

(8)

where αexp is the experimental conversion and αteo is the theoretical conversion, determined from Equation (2) and the kinetic triplet calculated from the reaction models, and N is the number of experimental TG data.

Table 2. Integral [a] and differential [b] form of various reaction models for solid phase kinetics. Information taken from Rueda-Ordoñez [52].

Reaction Model	Code	$f(\alpha)$ [a]	$g(\alpha)$ [b]
Power law	P1	$4\alpha^{3/4}$	$\alpha^{1/4}$
Power law	P2	$3\alpha^{2/3}$	$\alpha^{1/3}$
Power law	P3	$2\alpha^{1/2}$	$\alpha^{1/2}$
Power law	P4	$2/3\alpha^{1/2}$	$\alpha^{3/2}$
Phase-boundary controlled reaction (contracting area)	R2	$2(1-\alpha)^{1/2}$	$\left[1-(1-\alpha)^{1/2}\right]$
Phase-boundary controlled reaction (contracting volume)	R3	$3(1-\alpha)^{2/3}$	$\left[1-(1-\alpha)^{1/3}\right]$
Avrami Erofe'ev (m = 2)	A2	$2(1-\alpha)[-\ln(1-\alpha)]^{1/2}$	$[-\ln(1-\alpha)]^{1/2}$
Avrami Erofe'ev (m = 3)	A3	$3(1-\alpha)[-\ln(1-\alpha)]^{2/3}$	$[-\ln(1-\alpha)]^{1/3}$
Avrami Erofe'ev (m = 4)	A4	$4(1-\alpha)[-\ln(1-\alpha)]^{3/4}$	$[-\ln(1-\alpha)]^{1/4}$
First-order	F1	$(1-\alpha)$	$-\ln(1-\alpha)$
nth-order	Fn	$(1-\alpha)^n$	$\frac{[1-(1-\alpha)^{1-n}]}{1-n}$
One-dimensional diffusion	D1	$\frac{1}{2\alpha}$	α^2

Table 2. Cont.

Reaction Model	Code	$f(\alpha)$ [a]	$g(\alpha)$ [b]
Two-dimensional diffusion	D2	$[-\ln(1-\alpha)]^{-1}$	$(1-\alpha)\ln(1-\alpha)+\alpha$
Three-dimensional diffusion	D3	$\left(\frac{3}{2}\right)(1-\alpha)^{2/3}\left[1-(1-\alpha)^{1/3}\right]^{-1}$	$\left[1-(1-\alpha)^{1/3}\right]^{2}$
Ginstling–Brounstein diffusion	D4	$\left(\frac{3}{2}\right)\left[(1-\alpha)^{1/3}-1\right]^{-1}$	$\left(1-\frac{2\alpha}{3}\right)-(1-\alpha)^{2/3}$

3. Results and Discussion

3.1. Thermal Conduction

Figure 4 shows the influence of the density of the fique samples on their thermal conductivity. Density was one of the manufacturing parameters controlled during the elaboration of the samples, obtaining three levels of density: 50 kg/m^3, 65 kg/m^3, and 80 kg/m^3. As can be seen, there is a small decrease in thermal conductivity with increasing density in the 50 to 80 kg/m^3 range, where the maximum average value is 0.06 W/m·K at a density of 50 kg/m^3, while the minimum average value is 0.055 W/m·K at 80 kg/m^3. It seems reasonable to expect an increase in thermal conductivity with the increasing density of the nonwovens, as the solid fraction increases. However, the sample of fique nonwovens with the highest density has the lowest thermal conductivity value. This indicates that voids are also a parameter to be considered, as the results suggest that the higher the packing density, the smaller the size of the voids and the higher the thermal resistance.

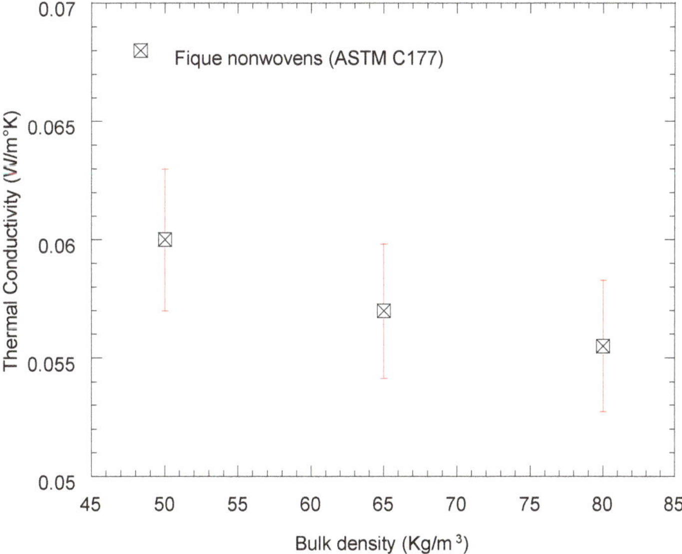

Figure 4. Results of thermal conductivity measurements at different densities.

3.2. Thermal Decomposition

Figure 5 shows the TG–DTG profile of the analyzed sample. As can be seen, there are five stages: a first stage of weight loss due to drying, a stable phase with no weight loss, two stages of high loss due to the decomposition process, and a final stage of degradation and combustion. It has been observed that hemicellulose, cellulose, and lignin decompose from 197 to 327 °C, 277 to 427 °C, and 277 to 527 °C, respectively [53]; therefore, the third stage can be attributed to the decomposition of hemicellulose and cellulose, while the fourth stage can be attributed to the decomposition of cellulose and lignin.

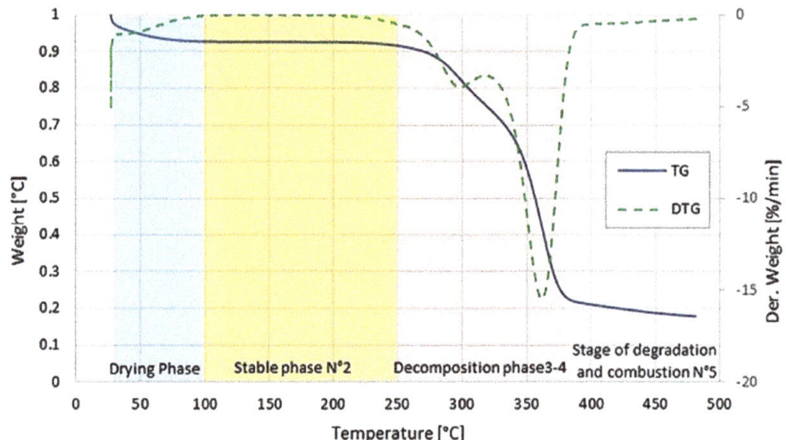

Figure 5. TG–DTG diagram of the fique samples.

Applying the Coats–Redfern method to the data obtained from the thermogravimetric tests, we obtained the straight line $\ln\left(\frac{g(\alpha)}{T^2}\right)$ vs. $\frac{1}{T}$ presented in Figure 6a. The correlation coefficient (R2) of this plot is 0.9903, which speaks well for the efficiency of the method. The values of the activation energy and pre-exponential factor obtained from this graph were 120.12 kJ/mol and 2.62×10^7. On the other hand, by applying the "Master Plots" method, it was found that the reaction model that best fit the analyzed biomass was the "Power law-P4", as presented in Figure 6b.

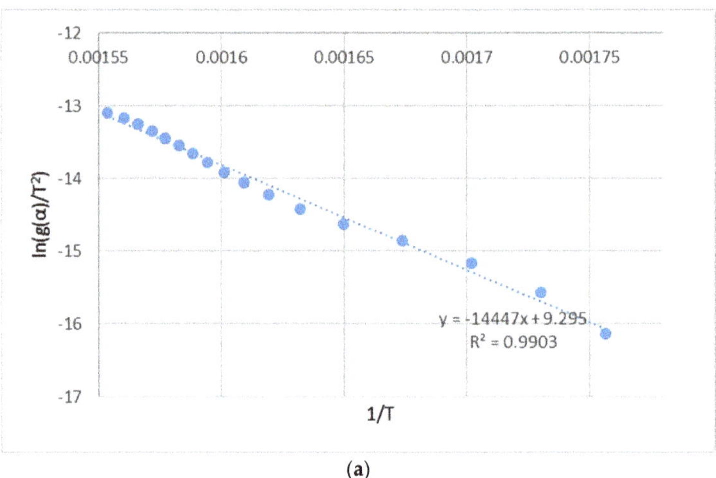

(a)

Figure 6. *Cont.*

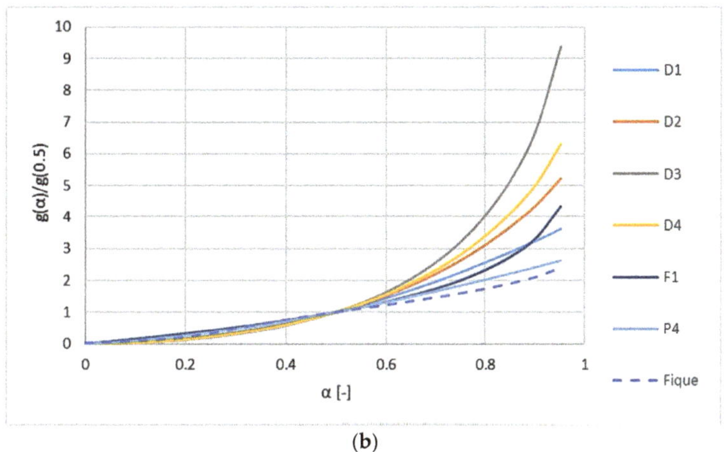

(b)

Figure 6. Determination of the kinetic triplet: (**a**) activation energy and pre-exponential factor by the Coats–Redfern method and (**b**) reaction model by means of the "Master Plots" method.

The results obtained show that, as expected, fique fibers decompose much faster than common insulators, so it is necessary to apply treatments or mix them with other materials to improve their resistance to high temperatures. The evaluation of the results by comparing the theoretical and experimental conversion curves is presented in Figure 7. As can be seen at the beginning and end of the process, there is a small underestimation of the results when using the model; however, the difference is very small, so a good fit is observed. To evaluate this adjustment, the average deviation percentage (AVP) was calculated, which gave a result of 3.7%, indicating that the results obtained can be used successfully in the simulation of the thermal decomposition of fique fibers.

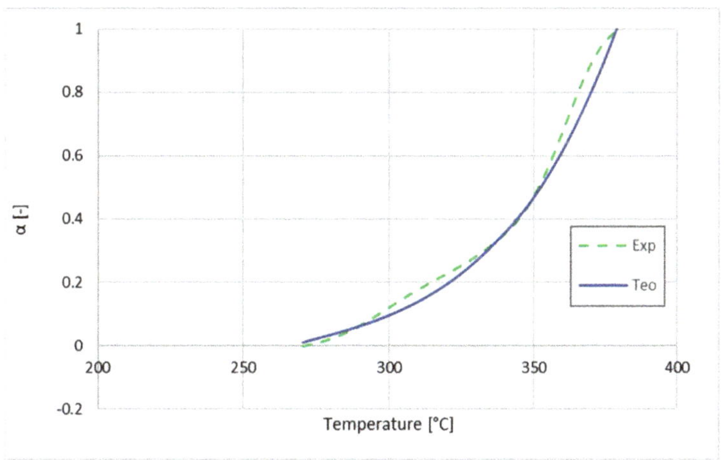

Figure 7. Theoretical and experimental conversion curves.

4. Conclusions

Equipment for measuring thermal conductivity was built, based on the ASTM C 177-13 standard, with which samples of nonwoven fique fibers of densities between 50 and 80 kg/m^3 were tested. The conductivity of the samples was found to be between 0.055 and 0.06 W/m·K, which is a value close to that of the common insulators found on the market.

25

This indicates that, from a heat transfer point of view, the material can be a good thermal insulator for buildings.

Thermogravimetric tests were also carried out in an inert atmosphere at 10 °C/min on the fique samples. It was observed that the thermal decomposition process consists of five stages: drying, heating without loss of mass, two stages of decomposition, and final degradation. These were carried out in the temperature ranges: 27–100 °C, 100–250 °C, 250–320 °C, 320–378 °C, and 378–480 °C, respectively.

From the TG data, the kinetic modelling of the biomass was carried out by means of the Coats–Redfern method, using the master plot curves for the determination of the reaction models, thus obtaining the kinetic triplet that allows us to model its thermal decomposition process. The results obtained were evaluated by comparing the theoretical and experimental conversion curves, obtaining average deviation percentages (AVP) of less than 4%, which leads us to the conclusion that the value of the kinetic triplet obtained is adequate for modelling the thermal decomposition process.

Very few studies evaluate the environmental impacts "cradle to grave" in a rigorous way, using, for instance, the life cycle assessment approach (LCA) of nonwoven fique samples. This lack of data is mainly caused by the state of the research on these materials that is still at an early stage. Further analyses should also be performed to evaluate other important properties for thermal insulators such as fire classification, resistance to water vapor diffusion, acoustic absorption, degradation due to moisture, bacteria, mildew, and fungi. In conclusion, issues remain to be solved before the widespread use of the fique nonwoven materials as thermal insulators.

Author Contributions: Conceptualization, G.F.G.S. and R.E.G.L.; methodology, G.F.G.S. and R.E.G.L.; validation, G.F.G.S., R.E.G.L. and R.A.G.-L.; resources, G.F.G.S., R.E.G.L. and R.A.G.-L.; review and editing G.F.G.S., R.E.G.L. and R.A.G.-L. All authors have read and agreed to the published version of the manuscript.

Funding: The authors wish to thank CEU San Pablo University Foundation for the funds dedicated to the Project Ref. USP CEU-CP20V12 provided by CEU San Pablo University.

Institutional Review Board Statement: Not applicable.

Informed Consent Statement: Not applicable.

Data Availability Statement: Not applicable.

Acknowledgments: Funding from Minciencias, Ministry of National Education, Ministry of Industry, Commerce and Tourism and ICETEX, Scientific Ecosystem—Colombia Científica, Fund Francisco José de Caldas, through Contract RC-FP44842-212-2018, is gratefully acknowledged and the Universidad Pontificia Bolivariana under grant No. 028-0717-2600.

Conflicts of Interest: The authors declare no conflict of interest.

References

1. Ortiz, A.C.; Peña, J.Z. Implicaciones Históricas y Biológicas del Cambio Climático. NOVA 2010, 8. Available online: http://unicolmayor.edu.co/publicaciones/index.php/nova/article/view/154 (accessed on 8 October 2015).
2. IPCC—Intergovernmental Panel on Climate Change. Cambio Climático 2014—Impactos, Adaptación y Vulnerabilidad. Available online: http://www.ipcc.ch/home_languages_main_spanish.shtml (accessed on 24 October 2016).
3. Liu, L.; Li, H.; Lazzaretto, A.; Manente, G.; Tong, C.; Liu, Q.; Li, N. The development history and prospects of biomass-based insulation materials for buildings. *Renew. Sustain. Energy Rev.* **2017**, *69*, 912–932. [CrossRef]
4. United Nations Environment Programme (UNEP)—SBCI. Available online: http://www.unep.org/sbci/AboutSBCI/Background.asp (accessed on 19 October 2016).
5. Buildings Energy Data Book. Available online: http://buildingsdatabook.eren.doe.gov/ChapterIntro1.aspx (accessed on 19 October 2016).
6. Buildings—Energy—European Commission. Energy. Available online: http://ec.europa.eu/energy/en/topics/energy-efficiency/buildings (accessed on 19 October 2016).
7. De Boeck, L.; Verbeke, S.; Audenaert, A.; De Mesmaeker, L. Improving the energy performance of residential buildings: A literature review. *Renew. Sustain. Energy Rev.* **2015**, *52*, 960–975. [CrossRef]

8. Carpio, C.; Coviello, M. Eficiencia Energética en América Latina y el Caribe: Avances y Desafíos del último Quinquenio. Available online: http://www.cepal.org/es/publicaciones/4106-eficiencia-energetica-america-latina-caribe-avances-desafios-ultimo-quinquenio (accessed on 19 October 2016).

9. Schiavoni, S.; D'Alessandro, F.; Bianchi, F.; Asdrubali, F. Insulation materials for the building sector: A review and comparative analysis. *Renew. Sustain. Energy Rev.* **2016**, *62*, 988–1011. [CrossRef]

10. Tangjuank, S. Thermal insulation and physical properties of particleboards from pineapple leaves. *Int. J. Phys. Sci.* **2011**, *6*, 4528–4532.

11. Papadopoulos, A. State of the art in thermal insulation materials and aims for future developments. *Energy Build.* **2005**, *37*, 77–86. [CrossRef]

12. Asdrubali, F.; D'Alessandro, F.; Schiavoni, S. A review of unconventional sustainable building insulation materials. *Sustain. Mater. Technol.* **2015**, *4*, 1–17. [CrossRef]

13. Binici, H.; Aksogan, O.; Demirhan, C. Mechanical, thermal and acoustical characterizations of an insulation composite made of bio-based materials. *Sustain. Cities Soc.* **2016**, *20*, 17–26. [CrossRef]

14. Mati-Baouche, N.; De Baynast, H.; Lebert, A.; Sun, S.; Lopez-Mingo, C.J.S.; Leclaire, P.; Michaud, P. Mechanical, thermal and acoustical characterizations of an insulating bio-based composite made from sunflower stalks particles and chitosan. *Ind. Crops Prod.* **2014**, *58*, 244–250. [CrossRef]

15. Mounika, M.; Ramaniah, K.; Prasad, A.R.R.; Rao, K.M.; Reddy, K. Thermal conductivity characterization of bamboo fiber reinforced polyester composite. *J. Mater. Environ. Sci.* **2012**, *3*, 1109–1116.

16. Binici, H.; Eken, M.; Dolaz, M.; Aksogan, O.; Kara, M. An environmentally friendly thermal insulation material from sunflower stalk, textile waste and stubble fibres. *Constr. Build. Mater.* **2014**, *51*, 24–33. [CrossRef]

17. Pinto, J.; Paiva, A.; Varum, H.; Costa, A.; Cruz, D.; Pereira, S.; Fernandes, L.; Tavares, P.; Agarwal, J. Corn's cob as a potential ecological thermal insulation material. *Energy Build.* **2011**, *43*, 1985–1990. [CrossRef]

18. Wei, K.; Lv, C.; Chen, M.; Zhou, X.; Dai, Z.; Shen, D. Development and performance evaluation of a new thermal insulation material from rice straw using high frequency hot-pressing. *Energy Build.* **2015**, *87*, 116–122. [CrossRef]

19. Hurtado, P.L.; Rouilly, A.; Vandenbossche, V.; Raynaud, C. A review on the properties of cellulose fibre insulation. *Build. Environ.* **2016**, *96*, 170–177. [CrossRef]

20. Madurwar, M.V.; Ralegaonkar, R.V.; Mandavgane, S.A. Application of agro-waste for sustainable construction materials: A review. *Constr. Build. Mater.* **2013**, *38*, 872–878. [CrossRef]

21. Ingrao, C.; Giudice, A.L.; Bacenetti, J.; Tricase, C.; Dotelli, G.; Fiala, M.; Siracusa, V.; Mbohwa, C. Energy and environmental assessment of industrial hemp for building applications: A review. *Renew. Sustain. Energy Rev.* **2015**, *51*, 29–42. [CrossRef]

22. Kymäläinen, H.-R.; Sjöberg, A.-M. Flax and hemp fibres as raw materials for thermal insulations. *Build. Environ.* **2008**, *43*, 1261–1269. [CrossRef]

23. Guzmán, R.E.; Gómez, S.; Amelines, O.; Aparicio, G.M. Superficial modification by alkalization of cellulose Fibres obtained from Fique leaf. In Proceedings of the IOP Conference Series: Materials Science and Engineering, Bristol, UK, 9–13 July 2018; Volume 437, p. 012015. [CrossRef]

24. An, L.; Wang, J.; Petit, D.; Armstrong, J.N.; Li, C.; Hu, Y.; Huang, Y.; Shao, Z.; Ren, S. A scalable crosslinked fiberglass-aerogel thermal insulation composite. *Appl. Mater. Today* **2020**, *21*, 100843. [CrossRef]

25. Dou, L.; Zhang, X.; Cheng, X.; Ma, Z.; Wang, X.; Si, Y.; Yu, J.; Ding, B. Hierarchical Cellular Structured Ceramic Nanofibrous Aerogels with Temperature-Invariant Superelasticity for Thermal Insulation. *ACS Appl. Mater. Interfaces* **2019**, *11*, 29056–29064. [CrossRef]

26. Vergara, M.C.A.; Gómez, M.P.C.; Restrepo, M.C.R.; Henao, J.M.; Soto, M.A.P.; Rojo, P.F.G.; Herazo, C.I.C.; Gallego, R.Z. Novel Biobased Textile Fiber from Colombian Agro-Industrial Waste Fiber. *Molecules* **2018**, *23*, 2640. [CrossRef]

27. Monteiro, S.; De Assis, F.S.; Ferreira, C.L.; Simonassi, N.T.; Weber, R.P.; Oliveira, M.S.; Colorado, H.A.; Pereira, A.C. Fique Fabric: A Promising Reinforcement for Polymer Composites. *Polymers* **2018**, *10*, 246. [CrossRef]

28. Fokaides, P.; Apanaviciene, R.; Černeckiene, J.; Jurelionis, A.; Klumbyte, E.; Kriauciunaite-Neklejonoviene, V.; Pupeikis, D.; Rekus, D.; Sadauskiene, J.; Seduikyte, L.; et al. Research Challenges and Advancements in the Field of Sustainable Energy Technologies in the Built Environment. *Sustainability* **2020**, *12*, 8417. [CrossRef]

29. Mansor, M.R.; Salit, M.S.; Zainudin, E.S.; Aziz, N.A.; Ariff, H. Life Cycle Assessment of Natural Fiber Polymer Composites. In *Agricultural Biomass Based Potential Materials*; Springer: Berlin/Heidelberg, Germany, 2015; pp. 121–141.

30. Cabeza, L.F.; Rincón, L.; Vilariño, V.; Pérez, G.; Castell, A. Life cycle assessment (LCA) and life cycle energy analysis (LCEA) of buildings and the building sector: A review. *Renew. Sustain. Energy Rev.* **2014**, *29*, 394–416. [CrossRef]

31. Muñoz, D.M.; Cifuentes, G.C. El fique como aislante térmico. *Biotecnol. Sect. Agropecu. Agroind.* **2007**, *5*, 9–16.

32. Onésippe, C.; Toro, F.; Bilba, K.; Delvasto, S.; Arsène, M.-A. Influence of Fibers Weight Fraction and Nature of Fibers on Thermal and Mechanical Properties of Vegetable Fibers/Cement Composites. In Proceedings of the 11th Conference on Durability of Building Materials and Components (11DBMC), Istanbul, Turkey, 11–14 May 2008. Available online: https://hal.univ-antilles.fr/hal-01693195/document (accessed on 18 June 2018.).

33. Monsalve, L.V.; Bolañoz, I.H.; Lopez, P.F.; Toro, E.F. Teja tipo sándwich de cemento basados en subproductos industriales para el mejoramiento de la comodidad térmica. *Rev. Colomb. Mater.* **2014**, *5*, 332–337.

34. Navacerrada, M.A.; Diaz, C.; Pedrero, A.; Fernández-Morales, P.; Navarro, G.; Cardona, O. Caracterización AcúStica de Muestras de Fique Tejido y no Tejido. Available online: http://www.sea-acustica.es/fileadmin/publicaciones/MAT-0_002.pdf (accessed on 12 May 2021).
35. Navacerrada, M.A.; Díaz, C.; Fernández, P. Characterization of a Material Based on Short Natural Fique Fibers. *BioResources* **2014**, *9*, 3480–3496. [CrossRef]
36. Proaño, E.; Bonilla, O.; Aldas, M. Desarrollo de un Material Compuesto de Matriz de Poliuretano Rígido Reforzado con Fibra de Raquis de Palma Africana. *Rev. Politécnica* **2015**, *36*, 89.
37. Navacerrada, M.A.; Díaz, C.; Pedrero, A.; Isaza, M.; Fernández, P.; Álvarez-Lopez, C.; Restrepo-Osorio, A. *Caracterización Acústica y Térmica de no Tejidos Basados en Fibras Naturales*; European Acoustics Association: Porto, Portugal, 2016.
38. Gutiérrez Vera, C.K. *Estudio del Rendimiento Acústico y Térmico de la Fibra de Cabuya Como Panel para Revestimiento de Paredes*; Universidad Laica Vicente Rocafuente: Guayaquil, Ecuador, 2017. Available online: http://repositorio.ulvr.edu.ec/handle/44000/2077 (accessed on 18 June 2018).
39. Sánchez, G.F.G.; Lopez, R.E.G.; Restrepo-Osorio, A.; Arroyo, E.H. Fique as thermal insulation morphologic and thermal characterization of fique fibers. *Cogent Eng.* **2019**, *6*, 1579427. [CrossRef]
40. Gomez, T.S.; Navacerrada, M.; Díaz, C.; Fernández-Morales, P. Fique fibres as a sustainable material for thermoacoustic conditioning. *Appl. Acoust.* **2020**, *164*, 107240. [CrossRef]
41. Hadded, A.; Benltoufa, S.; Fayala, F.; Jemni, A. Thermo physical characterisation of recycled textile materials used for building insulating. *J. Build. Eng.* **2016**, *5*, 34–40. [CrossRef]
42. ASTM C177—13 Standard Test Method for Steady-State Heat Flux Measurements and Thermal Transmission Properties by Means of the Guarded-Hot-Plate Apparatus. Available online: https://www.astm.org/DATABASE.CART/HISTORICAL/C177-13.htm (accessed on 16 May 2021).
43. Sher, F.; Iqbal, S.Z.; Liu, H.; Imran, M.; Snape, C.E. Thermal and kinetic analysis of diverse biomass fuels under different reaction environment: A way forward to renewable energy sources. *Energy Convers. Manag.* **2020**, *203*, 112266. [CrossRef]
44. Slopiecka, K.; Bartocci, P.; Fantozzi, F. Thermogravimetric analysis and kinetic study of poplar wood pyrolysis. *Appl. Energy* **2012**, *97*, 491–497. [CrossRef]
45. Jones, J.; Saddawi, A.; Dooley, B.; Mitchell, E.; Werner, J.; Waldron, D.; Weatherstone, S.; Williams, A. Low temperature ignition of biomass. *Fuel Process. Technol.* **2015**, *134*, 372–377. [CrossRef]
46. Coats, A.W.; Redfern, J.P. Kinetic Parameters from Thermogravimetric Data. *Nat. Cell Biol.* **1964**, *201*, 68–69. [CrossRef]
47. Álvarez, A.; Pizarro, C.; García, R.; Bueno, J.; Lavín, A. Determination of kinetic parameters for biomass combustion. *Bioresour. Technol.* **2016**, *216*, 36–43. [CrossRef]
48. Rueda-Ordóñez, Y.J.; Tannous, K. Isoconversional kinetic study of the thermal decomposition of sugarcane straw for thermal conversion processes. *Bioresour. Technol.* **2015**, *196*, 136–144. [CrossRef]
49. Vyazovkin, S.; Burnham, A.K.; Criado, J.M.; Perez-Maqueda, L.A.; Popescu, C.; Sbirrazzuoli, N. ICTAC Kinetics Committee recommendations for performing kinetic computations on thermal analysis data. *Thermochim. Acta* **2005**, *520*, 1–19. [CrossRef]
50. Perez-Maqueda, L.A.; Criado, J.M. The Accuracy of Senum and Yang's Approximations to the Arrhenius Integral. *J. Therm. Anal. Calorim.* **2000**, *60*, 909–915. [CrossRef]
51. Órfão, J.; Antunes, F.; Figueiredo, J. Pyrolysis kinetics of lignocellulosic materials—Three independent reactions model. *Fuel* **1999**, *78*, 349–358. [CrossRef]
52. Rueda-Ordoñez, Y.J. Thermal Decomposition Analysis of Sugarcane Straw in Inert and Oxidative Atmospheres through Thermo-analytical Methods. Ph.D. Thesis, University of Campinas, Campinas, Brazil, 2016. Available online: http://repositorio.unicamp.br/bitstream/REPOSIP/320795/1/RuedaOrdonez_YesidJavier_D.pdf (accessed on 2 September 2019).
53. Du, Z.; Sarofim, A.F.; Longwell, J.P. Activation energy distribution in temperature-programmed desorption: Modeling and application to the soot oxygen system. *Energy Fuels* **1990**, *4*, 296–302. [CrossRef]

Article

Modelling Long-Term Urban Temperatures with Less Training Data: A Comparative Study Using Neural Networks in the City of Madrid

Miguel Núñez-Peiró [1,*], Anna Mavrogianni [2], Phil Symonds [2], Carmen Sánchez-Guevara Sánchez [1] and F. Javier Neila González [1]

[1] School of Architecture, Universidad Politécnica de Madrid, Avda. Juan de Herrera 4, 28040 Madrid, Spain; carmen.sanchezguevara@upm.es (C.S.-G.S.); fjavier.neila@upm.es (F.J.N.G.)

[2] Institute of Environmental Design and Engineering, University College London, Central House, 14 Woburn Place, London WC1H 0NN, UK; a.mavrogianni@ucl.ac.uk (A.M.); p.symonds@ucl.ac.uk (P.S.)

* Correspondence: miguel.nunez@upm.es

Abstract: In the last decades, urban climate researchers have highlighted the need for a reliable provision of meteorological data in the local urban context. Several efforts have been made in this direction using Artificial Neural Networks (ANN), demonstrating that they are an accurate alternative to numerical approaches when modelling large time series. However, existing approaches are varied, and it is unclear how much data are needed to train them. This study explores whether the need for training data can be reduced without overly compromising model accuracy, and if model reliability can be increased by selecting the UHI intensity as the main model output instead of air temperature. These two approaches were compared using a common ANN configuration and under different data availability scenarios. Results show that reducing the training dataset from 12 to 9 or even 6 months would still produce reliable results, particularly if the UHI intensity is used. The latter proved to be more effective than the temperature approach under most training scenarios, with an average RMSE improvement of 16.4% when using only 3 months of data. These findings have important implications for urban climate research as they can potentially reduce the duration and cost of field measurement campaigns.

Keywords: urban heat island; microclimate; feed-forward neural networks; air temperature measurements; in-situ measurements; urban models; urban environment; climate change

Citation: Núñez-Peiró, M.; Mavrogianni, A.; Symonds, P.; Sánchez-Guevara Sánchez, C.; Neila González, F.J. Modelling Long-Term Urban Temperatures with Less Training Data: A Comparative Study Using Neural Networks in the City of Madrid. *Sustainability* **2021**, *13*, 8143. https://doi.org/10.3390/su13158143

Academic Editor: Roberto Alonso González Lezcano

Received: 17 June 2021
Accepted: 13 July 2021
Published: 21 July 2021

Publisher's Note: MDPI stays neutral with regard to jurisdictional claims in published maps and institutional affiliations.

1. Introduction

In the context of raising awareness on climate change, a good understanding of urban climate phenomena is a key milestone in order to mitigate and adapt to thermal extremes within urban environments [1,2]. Cities are not only one of the main contributors to the greenhouse effect [3], but also places where many inequalities and therefore potential vulnerabilities accumulate [4–6]. Moreover, recent studies, such as those developed by Grimm et al. [7] and Youngsteadt [8], suggest that cities could provide important insights into the socio-ecological dynamics of our near future at a global scale, thus increasing the interest for reliable urban climatic data and expanding its applications to many other disciplines.

However, obtaining reliable climatic data within urban areas is still a challenging task due to the complexity of the urban climate. Nowadays, some of the most important advances concentrate on the modelling field [9]. Examples can be found evaluating the inter-relation between some parameters and the urban climate, such as the presence of water-bodies [10] or the emission of anthropogenic heat [11,12]. Regarding the accuracy of these numerical models, recent advances coupling urban canopy models with meso-climatic ones have also proved their overall reliability [13,14]. However, there are still some barriers that limit their applications in other fields. For example, Computational Fluid

29

Dynamics (CFD) has proved to be reliable for both building-scale models and relatively small urban areas (i.e., within a few hundred meters, [15,16]) but too computing intensive for larger domains [17,18]. Other authors, such as a Lauzet et al. [19], have highlighted that the high computational needs of high-resolution urban climate models pose a significant challenge in obtaining long-term datasets, therefore hindering their more widespread use of urban models within building energy simulations.

Conducting on-site measurement campaigns is also one of the most widespread practices towards improving urban climate knowledge [20,21]. They are still an essential component of numerical model validation processes [22]. Regarding meteorological parameters, experimental data is primarily derived from urban networks consisting of multiple sensors distributed across the city [23]. Several examples can be found in the literature for cities all around the globe, such as in Athens [24], London [25,26], Sendai [27], Szeged [28], Guangzhou [29], Kaohsiung [30], Guwahati [31], Augsburg [32], Nanjing [33,34], Rotterdam [35] or Berlin [36]. However, these urban networks are expensive to deploy and maintain, thus their use is usually constrained in time and space, limiting their suitability in long-term studies.

Other sources of experimental data might also present important drawbacks. Citizen Weather Stations (CWS) have grown exponentially in recent years [37,38] and are being used in a variety of ways, from studying the intra-urban temperature patterns [39] to complementing weather forecasts [40]. However, they require sophisticated filtering techniques and quality control procedures to manage their calibration bias, instrument errors and representativeness issues [41,42]. Mobile measurements, another widely adopted practice to study the spatial distribution of the UHI in detail, has expanded in recent years from the traditional approach of car transects [43–45] to bicycle transects [46–49] or even drone transects [50]. Despite their versatility, mobile measurements are very demanding in terms of human resources and can hardly be used to obtain time series at a fine scale (i.e., hourly). The latter is also one of the main drawbacks of remote sensing techniques, which depend on the timing of the satellite overpass, and require post-processing to address the presence of clouds and limited view angles [51].

1.1. Data-Driven Approaches for Modelling Outdoor Urban Temperatures

A widespread alternative technique for obtaining reliable and affordable long-term datasets of urban air temperatures is the development of empirical models. These models use pre-existing statistical correlations among available data to generate accurate projections without compromising their computational efficiency. Consequently, these data-driven approaches represent bespoke alternatives to more complex numerical models.

Several algorithms can be used for this purpose. A widely used technique for modelling urban temperatures is using Multiple Linear Regression (MLR), which has been tested for both temporal [52–55] and spatial predictions [56–60]. However, the increasing availability of machine learning and big data solutions is boosting the widespread use of other algorithms which, although potentially harder to interpret, are likely to improve their accuracy. Popular machine learning techniques include Support Vector Machines [61–64], Random Forest [58,60,62,65,66], or Artificial Neural Networks (ANN).

ANN seem to stand as the most popular approach for modelling the hourly evolution of outdoor urban temperatures. To the authors' knowledge, Mihalakakou et al. [67] presented the first attempt to model the outdoor temperature at an urban site using ANNs. They used the dry-bulb temperature data available from two existing meteorological stations in Athens: one located within the city (the target), and one at the outskirts (the reference site). In a follow-up study, the model was adapted for other urban sites in the same city, where they deployed a network of 23 temperature sensors across the city for 2 years [68,69].

In these early attempts to model urban temperatures using ANNs, the authors only used the air temperature from the reference site as the input. However, other researchers have explored the inclusion of additional predictors to increase model performance. The most common ones are meteorological parameters linked with the UHI formation. Kim and Baik [70], for example, used the maximum UHI intensity of the previous day in Seoul together with wind speed, cloud cover, and relative humidity. In London, Kolokotroni et al. [71–73] used hourly air temperature, relative humidity, wind speed, cloud cover and global solar radiation. More recently, in Ontario, Demirezen et al. [74,75] used the air temperature, humidity, solar radiation, wind speed and wind direction. Other researchers have also included a time reference as an input to better capture the hourly evolution of urban temperatures. For example, Gobakis et al. [24] and Papantoniou and Kolokotsa [76] used the date in conjunction with air temperature and global solar radiation. Similarly, Heijden et al. [35] and Erdemir and Ayata [77] used the hour of the day together with other meteorological parameters. Table 1 summarizes these and other ANN studies that focused on outdoor urban temperatures and their modelling characteristics, such as the length of their datasets.

Table 1. Previous studies using ANN to model the outdoor air temperature in urban areas, in chronological order.

Reference	City, Country [a]	Training and Testing Dataset			ANN Target [b]	ANN Type
		Initial Date	Final Date	Duration		
Mihalakakou et al. [67]	Athens, GR	1986	1995	10 years	Temperature	FNN
Santamouris et al. [69]	Athens, GR	Jun 1996 Jun 1997	Sep 1996 Sep 1997	8 months	Temperature	FNN
Kim and Baik [70]	Seoul, KR	1973	1996	24 years	UHI intensity	FNN
Mihalakakou et al. [68,78]	Athens, GR	Jan 1996	Dec 1998	2 years	UHI intensity	FNN
Jang et al. [79]	Québec [1], CA	Jun 2000	Sep 2000	4 months	Temperature	FNN
Kolokotroni et al. [71–73]	London, GB	Jul 1999 2007	Sep 2000 2007	15 months	Temperature	FNN, CNN, ENN
Zhao [80]	Quinling [1], CN	-	-	-	Temperature	FNN
Beccali et al. [81]; Cellura et al. [82]	Palermo, IT	-	-	-	Temperature	NNARX, NNARMAX
Gobakis et al. [24]	Athens, GR	Apr 2009	May 2010	13 months	Temperature	FNN, CNN, ENN
Shao et al. [83]	Hangzhou, CN	Jan 1995	Dec 1996	2 years	Temperature	FNN
Heijden et al. [35]	Rotterdam, NL	Apr 2011	Oct 2012	19 months	UHI intensity	FNN
Lee et al. [84]	Seoul, KR	Jan 2012	Dec 2012	1 year	UHI intensity	FNN
Papantoniou and Kolokotsa [76]	Ancona, IT Chania, GR Granada, ES Mollet, ES	Jan [3]	Dec [3]	1 year	Temperature	FNN, CNN, ENN
Erdemir and Ayata [77]	Istanbul [2], TR	May [3]	Sept [3]	5 months	Temperature	FNN
Schuch et al. [85]	Abu Dhabi, AE	Mar 2016	Dec 2016	10 months	Temperature	FNN
Demirezen et al. [74,75]	Ontario, CA	Feb 2018	Nov 2018	9 months	Temperature	FNN
Han et al. [86]	Cambridge, US	Jan, 2019	Jun, 2019	6 months	Temperature	FNN, RNN

[a] ISO Country codes [87]. [b] Output of the ANN model, as declared or shown by the authors. [1] Extends further from the limits of the city, covering the surrounding regional areas. [2] Includes other cities of the same country. [3] Year not specified.

In most of these studies, the modelling of outdoor urban air temperature time series is addressed from a common perspective: using the temperatures collected during a monitoring campaign at the urban level to train a Feed-forward Neural Network (FNN, a relatively simple type of ANN). This modelling is usually performed using data from one or several reference points, in many cases well-established meteorological observatories providing detailed and robust information on a wide range of parameters. Although this process is quite extended, it could be discussed whether other ANN topologies might be more suitable for this purpose. Cascade Neural Networks (CNN) or Elman Neural Networks (ENN) have also been widely applied [24,72,76], the latter being simplified versions of Recurrent Neural Networks (RNN). RNNs have proved to be very effective when it comes to make forecasts, especially when Long Short-Term Memory (LSTM) is used [88]. In that sense, the work of Han et al. [86] has recently demonstrated the superiority of RNNs over FNNs for predicting outdoor urban temperatures.

However, it should be noted that the aim of most of these studies is not to make time predictions or forecasts, but to model an urban time series from a preexisting one. In other words, the purpose is to obtain an adapted version of a reference time series that already exists, being this new time series representative of a certain urban area and covering the exact same period as the data used as a reference. This simplifies the process by eliminating the time dependence of the outputs, and which justifies working with simpler neural networks, such as FNNs. In fact, and under this modelling scenario, Kolokotroni et al. [72] did not find any improvement when comparing ENNs and CNNs with FNNs.

Although empirical models are site-specific (predictions are always made for a particular urban location), they can be used to extend the temporal coverage of urban monitoring campaigns, thus potentially increasing their utility among other disciplines. And despite FNN-based models are not suitable for future projections, they are certainly useful to adapt historical records obtained outside the city to the reality of urban areas. However, there is currently a knowledge gap with regard to the amount of input data potentially needed to accurately model urban temperature time series using FNNs. Collecting experimental data is very time-consuming and resource-intensive and, while it seems a common practice to rely on one whole year of data for the training, there is no evidence that this should be a minimum requirement. This study, therefore, aims to quantify the degree to which the amount of input data needed to train FNNs can be reduced without sacrificing their accuracy. We also explore the use of the UHI intensity as an alternative output of the FNN models, instead of directly targeting the air temperature, to test the hypothesis that its lower seasonality and direct association with the input variables might help reduce the amount of required data for the training phase.

The present research is structured in three phases: first, we compared the performance of more than 5000 different FNN configurations for modelling the outdoor urban temperature (TEMP approach) and the UHI intensity (UHII approach) when trained with 12 months of data in the city of Madrid. An optimal configuration was then selected and analysed further in-depth for both approaches, including their sensitivity to input parameters. Finally, the amount of data provided during the training phase was reduced from the initial 12 months to 9, 6 and 3 months to evaluate the capacity of these models to continue producing accurate results with fewer input data.

2. Materials and Methods

2.1. Study Area: The City of Madrid

The present study focuses on the city of Madrid. Due to its size, location and climatic conditions, Madrid is characterised by a strong UHI, with nighttime UHI intensities up to 10 °C during calm and clear nights. During the last decades, this phenomenon has been intensively studied in the city by means of on-site measurements [89–92], remote sensing [93,94] and numerical models [95,96].

Between 2016 and 2019, a continuous monitoring campaign was carried out at 20 fixed urban sites with the aim to study the temporal patterns of the UHI in Madrid [97]. In the present study, we use part of that experimental data to define the outputs of our ANN models. More specifically, we use the hourly, dry-bulb temperature gathered at the city centre (Embajadores, see Figure 1), classified as compact midrise (LCZ 2) according to the Local Climate Zones (LCZ) scheme [98], and which registered the highest mean and nighttime UHI intensity. The data available for this study cover the period from July 2016 to September 2018 on an hourly basis (800 days or 19,200 h, in total).

All sensors used in this monitoring campaign were protected from the rain and solar radiation using a custom-made, mechanically ventilated radiation shield. They were installed in the Urban Canopy Layer (UCL) at 5–6 m above the ground, following the guidelines of the World Meteorological Organization (WMO) for urban sites [99,100]. The location of each sensor was also studied in terms of its thermal source area [101]. In that sense, the representativeness of each sensor was appraised in terms of its surroundings' homogeneity [102,103].

Quality Control (QC) procedures were also applied, consisting of a plausible value check, a time consistency check, and an internal-consistency check [104]. This analysis was complemented by a spatial consistency check [105], which analysed whether the difference between a measurement and its surroundings was too large compared to the average. For the City Centre sensor, 126 records were flagged as suspect and just three as erroneous. 72 missing values were identified due to a recording failure between the 17th and the 20th of October 2017. Both erroneous and missing values were left blank in the analysed dataset. Further details about the monitoring campaign and QC procedures can be found in [97].

In addition to the experimental data collected at the city centre, records from the nearby meteorological stations of Barajas Airport (LCZ D) and Ciudad Universitaria (LCZ 9) were used. Hourly values of dry bulb temperature, relative humidity, wind speed, wind direction and precipitation were extracted from the former, while global solar radiation was obtained from the latter. The data covered the same time period (July 2016–September 2018). Both stations are managed by the Spanish Meteorological Agency (AEMET), which complies with the requirements established by the WMO Integrated Global Observing System (WIGOS, [106,107]) regarding QC and sensor installation.

Three different types of datasets, the training, validation and the test datasets, were created. The former were used to fit and evaluate different ANN model configurations. Several training and validation datasets, which varied in length (12, 9, 6 and 3 months) and the months that they covered, were created based on almost 15 months of monitoring (July 2016–September 2017, 10,440 records/hourly measurements). All these datasets were continuous over time, and they were distributed as 80% training and 20% validation. These training and validation subsets were created by randomly sampling the data. This prevented the potential accumulation of specific events in any of these datasets (e.g., certain meteorological conditions), which could bias either the training or the validation of the models. Additionally, a test dataset was created based on the second year of recorded data (October 2017–September 2018, 8688 records/hourly measurements) to independently test the models and assess their accuracy over an entirely different year.

2.2. Designing the ANNs

Feed-forward Neural Networks (FNN) were used in this study. Although FNNs are at the baseline of supervised deep neural networks, their utility for modelling urban temperatures has been widely demonstrated in previous studies (see Section 1.1). Figure 2 outlines the two different approaches, based on two different outputs, that were adopted in this study to model urban temperatures. The first one consisted of directly targeting the air temperature at the urban site, validating its outputs with the measurements previously recorded at that location. This approach is aligned with the majority of similar studies found in the literature, and it is referred in this study as the temperature approach (TEMP approach). The second option aims at modelling the urban air temperature indirectly. In this case, the model targets the UHI intensity instead, computed as the temperature difference between the urban site (*Embajadores*) and the reference location (*Barajas Airport*, $\Delta T_{LCZ2, LCZD}$). The urban temperature is then derived indirectly by adding the airport temperature to the output of the model. This will be referred to as the UHII approach from this point onwards.

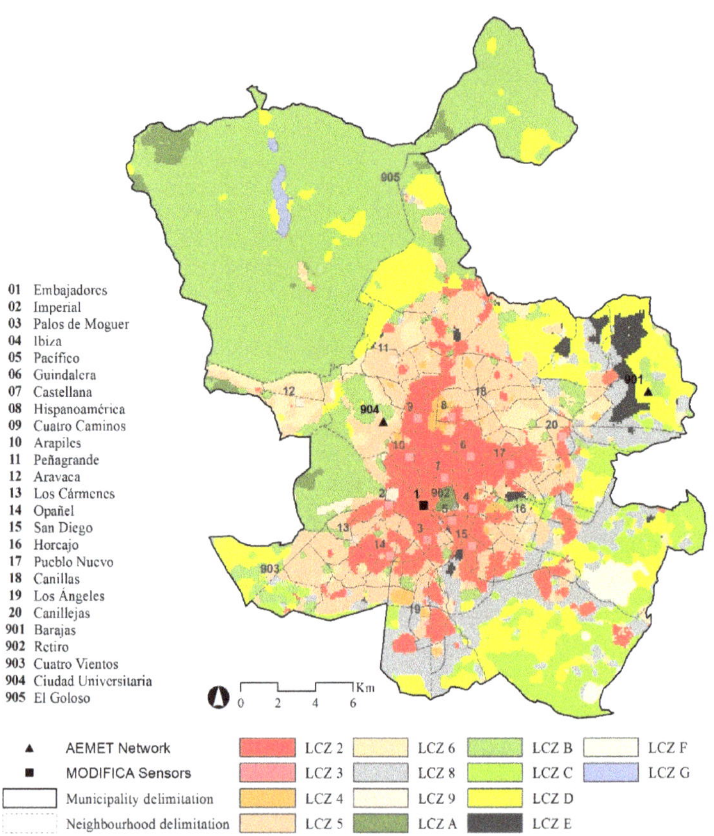

Figure 1. Distribution of the MODIFICA and AEMET networks across the city of Madrid. The data needed for the training, validation and test of the ANN model was extracted from the measurement sites in black. The classification of Madrid by Local Climate Zones, extracted from the WUDAPT database [108], is presented in the background.

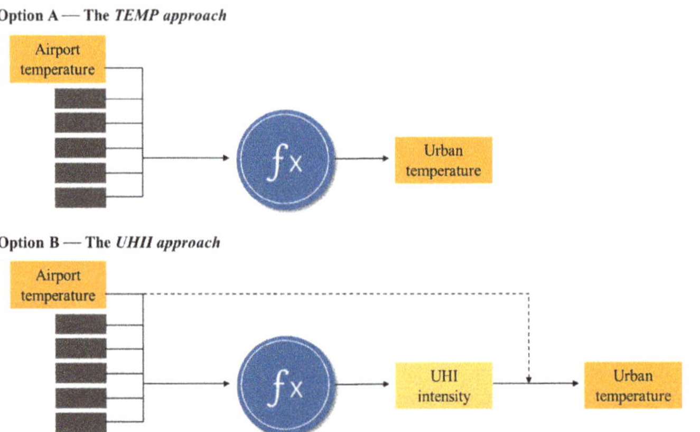

Figure 2. Schematic representation of the two approaches used for modelling the outdoor urban temperature.

The selection of the FNN model inputs of this study were informed by previous studies in Table 1, which have identified the variables that have a strong correlation with the formation of heat islands [109,110]. They consist of six meteorological variables: dry bulb temperature (°C), relative humidity (%), precipitation (mm), wind direction (degrees), wind speed (m/s) and global solar radiation (J/m^2). The time of the day was added to these six input parameters, which was expected to reflect the daily variability of the outputs, either the temperature or the UHI intensity. Cloud cover was not used as an input parameter because the available frequency (one record every eight hours) was incompatible with the hourly frequency for the outputs. The wind speed presented strong variations at an hourly level and introduced strong oscillations in the prediction. Thus, to help avoid abrupt changes in the output, a moving average (MA) filter was applied. The use of a MA filter is a common pre-processing technique when it comes to modelling time series from data with a high variability. Examples can be found in the field of urban traffic (applying a MA to the car's acceleration [111]), atmospheric pollution (MA applied to measured PM$_{2.5}$ concentration [112]) or urban climate modelling [113], the latter using a MA of order 8 (i.e., 8 h) to reduce the presence of wind gust peaks in the dataset prior feeding their model. In this study, a MA of order 4 (4 h) was found to be sufficient to reduce the noise of the wind speed while preserving the time series trend.

All the inputs were standardized prior the FNN feeding, meaning that all variables were transformed in order to have a mean = 0 and a standard deviation = 1 [114,115]. A diagram of the FNN structure for both approaches can be seen in Figure 3.

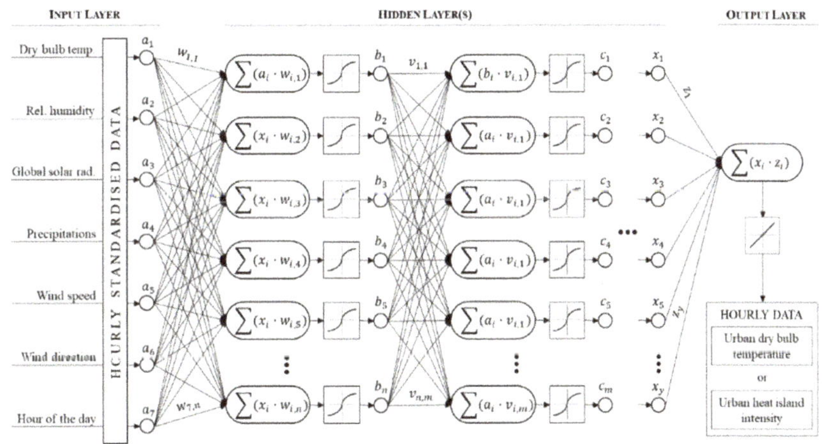

Figure 3. Base structure of the Feed-forward Neural Network (FNN) used in this study.

2.3. Comparing and Evaluating the FNNs

Several FNN structures with different configurations were trained during the first phase of this research. Hyperparameters, such as the number of neurons per hidden layer, the activation functions, or the number of epochs, were thoroughly iterated in order to find a common, optimal configuration for both the TEMP and the UHII approach. Despite some of the tested activation functions are commonly applied for classification tasks and were not likely to give the best performance (i.e., sigmoid-like functions), they were included in the iterative process since preceding similar works made use of them [24,67]. To streamline the process and reduce the complexity of the iteration, each subsequent hidden layer adopted half the neurons of the previous one. All models initialized their weights randomly and were initially trained using 12 months of data. Each configuration was compared by iterating just one parameter (e.g., the activation functions) and leaving the others fixed, while increasing the number of neurons per hidden layer. Those parameters that reached the best overall accuracy with the lowest number of neurons were selected. After this iterative process 5478 FNNs were trained. Table 2 summarizes the parameters used to test these configurations, as well as the ones that were finally selected. The task outlined above was performed using Python and Keras, a deep-learning library based on Tensorflow [116,117].

Once a common structure and configuration were defined, a comparative analysis of these models was carried out. First, the contribution of each input to the model output was assessed using a sensitivity analysis [114,118,119]. The 5th, 25th, 50th, 75th and 95th percentiles were used to run the sensitivity analysis for each input, while fixing the rest on their means. The time of the day was excluded from the sensitivity analysis and fixed at two different moments: noon and midnight. Next, their overall accuracy was compared for the TEMP and the UHII approach using several error metrics, such as the root mean squared error (RMSE), the median absolute deviation (MAD) or the coefficient of determination (R^2). Modelled results were then plotted for three different weeks to visually assess whether the modelling ability of any of these two approaches could be compromised under certain scenarios. These corresponded to a week of high atmospheric stability (and thus, strong UHI intensity), a week of high atmospheric instability (weak UHI intensity), and a week under both of these conditions.

Table 2. Parameters used to train and evaluate different FNNs configurations. It includes the configuration that was finally selected for both the temperature and the UHII approach.

Parameters		Tested	Selected
Number of hidden layers		1–5	2
Number of neurons	Input layer	7	7
	Hidden layers [1]	3–85	18
	Output layer	1	1
Activation functions	Hidden layers	Linear, ELU, SELU, ReLU, Sigmoid, Hard sigmoid, Hyperbolic tangent, Exponential, Softmax, Softplus, Softsign	ELU
	Output layers	Linear, ELU, SELU, ReLU, Sigmoid, Hard sigmoid, Hyperbolic tangent, Exponential, Softmax, Softplus, Softsign	Linear
Optimizer		SGD, Adam, RMSProp, Adagrad, Adadelta, Nadam	Adam
Epochs		100, 200, 500	200
Batch size		2, 5, 10	10
Dataset length		12 months	12 months [2]
Train/Validation size		80%/20%	80%/20%

[1] The value here presented corresponds to the number of neurons contained in the first hidden layer. Each subsequent hidden layer adopts half of the value of the previous one. [2] Maximum length of the dataset. Results with shorter lengths are also presented (Table 4).

The last step of the evaluation process consisted of modifying the amount of data provided to the neural networks during the training phase. To this end, FNN models for both the TEMP and the UHII approach were trained using 12, 9, 6 and 3 months of data, and were used to model the outdoor air temperatures for one complete year using the test dataset. The accuracy was estimated, as in the previous cases, using common error metrics. The loss of accuracy of the models trained with shorter datasets was addressed by comparing their performance with the models trained on more data, obtaining a percentage indicating the increase of error for each metric. In the case of models trained with just 3 months of data, the Mean Absolute Error (MAE) was estimated on a monthly basis to further explore its distribution along one year of modelling.

3. Results

A comparison between several FNN configurations is first shown in Figure 4. Each graph represents the overall accuracy of a certain FNN when iterating just one of its parameters, and while increasing the number of neurons in the hidden layers. From this iterative process, a common, optimal FNN configuration for both the TEMP and the UHII approach was established. The optimal structure was defined as a neural network with seven inputs, two hidden layers of 18 and 9 neurons respectively, and one output. In that sense, it was found that increasing from one to two hidden layers produced a significant improvement in the models' accuracy, while increasing the number of hidden layers further did not. Similarly, moving from 100 to 200 epochs during the training phase could reduce the error of the FNN, while the computational expense of using 500 epochs instead of 200 did not seem justified. This was particularly evident when having tens of neurons in the hidden layers.

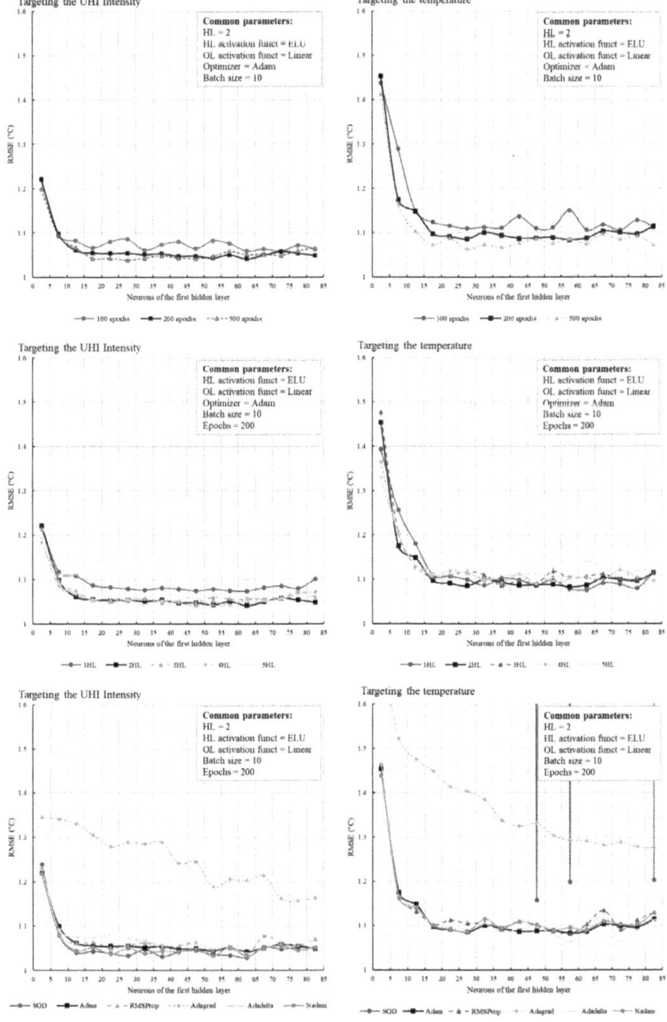

Figure 4. Average Root Mean Squared Error for different FNN configurations using Table 2. Results

obtained with the models derived from the TEMP and the UHII approach for the site Embajadores. Three weeks were selected, each one representing a different atmospheric stability scenario. The timeframe used to train these models extends from July 2016 to September 2017.

In some cases, due to the performance differences between the TEMP and the UHII approach, a common ground had to be reached in terms of the optimal configuration. That was the case of the activation functions, where the Stochastic Gradient Descent (SGD) seemed to produce the best results for those FNNs modelling the UHI intensity, but it led to exploding gradient problems when modelling the temperature. Thus, the Adaptive Moment Estimation (Adam) optimiser was used instead, which performed optimally in both scenarios. For the activation functions, a combination of the Exponential Linear Unit (ELU) for the hidden layers and the linear function for the output layer was used.

Overall, UHII models presented fewer converging problems than TEMP models, which seemed to have some difficulties with some activation functions and optimizers. Furthermore, the UHII approach usually outperformed the TEMP approach. The former did not only produce models with relatively smaller errors than the latter but required fewer neurons per hidden layer to reach a similar accuracy. This behaviour might be indicative of a clearer and more direct relationship between inputs and output, which in the case of the UHII approach links parameters such as wind speed, precipitation, or solar radiation with the UHI formation.

Differences between both modelling approaches also arise when looking at the inputs' relevance. In that sense, the sensitivity analyses presented in Figures 5 and 6 seem to reveal significant variations among them. The temperature from the reference site shifts from being the most relevant parameter of the entire FNN (TEMP approach) to being one of the least important (UHII approach). This is especially visible at night, when inter- and intra-urban temperature differences are most pronounced. The other parameters, albeit with different magnitude, seem to condition the outcome of both models in a similar way. In that sense, wind speed and direction seem to be two highly influential parameters during the night, while solar radiation and relative humidity seem to be key during the day.

Although the UHII approach appears to yield more balanced models, this apparent advantage does not seem to have a significant impact on their outputs when trained with 12 months of data. In this scenario, reasonably good results, and with similar error patterns, are obtained for both approaches. As it can be noted in Figure 7, modelled temperatures fit satisfactorily with the measured temperatures at the urban site and under a wide variety of circumstances, including different UHI scenarios: a rainy and windy week with generalised low UHI intensities (<2 °C); a week with varying meteorological conditions, during which a sudden weather change from calm to rainy was observed leading to a rapid change in the UHI intensities; or a calm week with strong UHI intensities (>5 °C), probably reinforced by temperature inversions. The greatest errors seem to accumulate on those nights when unusual conditions occur, such as when very high UHI intensities, close to 10 °C, are registered; or when the UHI intensity drops and rises abruptly, perhaps coinciding with occasional and localised weather events, such as rainfalls. Overall, models produced relatively smooth time series, without spikes or large variations from one hour to the next one, despite not having a built-in temporal dependence between consecutive outputs. Using a moving average for the wind speed seems to have contributed to reducing the noise in the models' output.

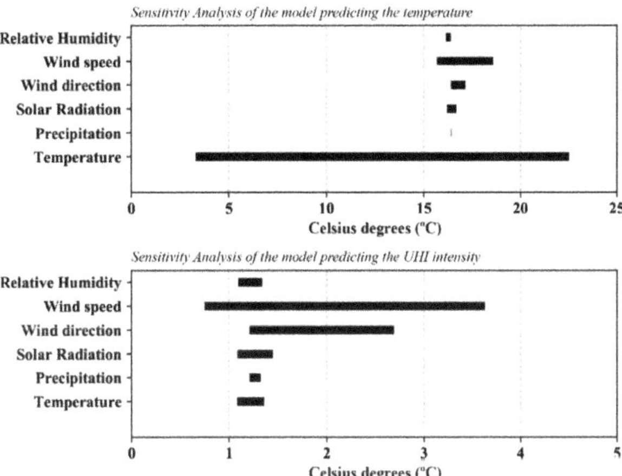

Figure 5. Sensitivity analysis of the inputs of the models. The same FNN configuration was used targeting the urban temperature (top) and the UHI intensity (bottom). The hour was fixed to 12 a.m. (Nighttime).

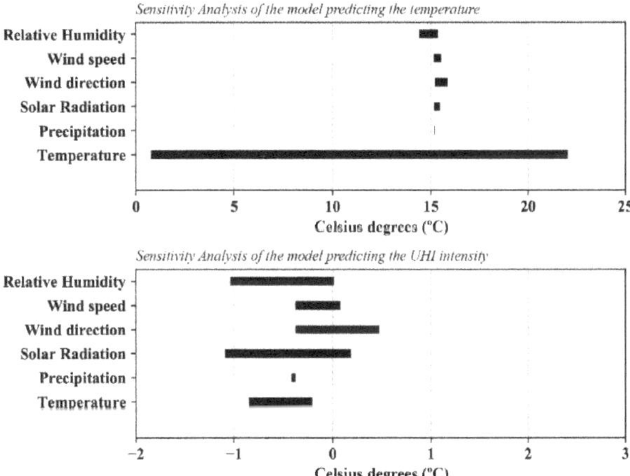

Figure 6. Sensitivity analysis of the inputs of the models. The same FNN configuration was used targeting the urban temperature (top) and the UHI intensity (bottom). The hour was fixed at 12 p.m. (Daytime).

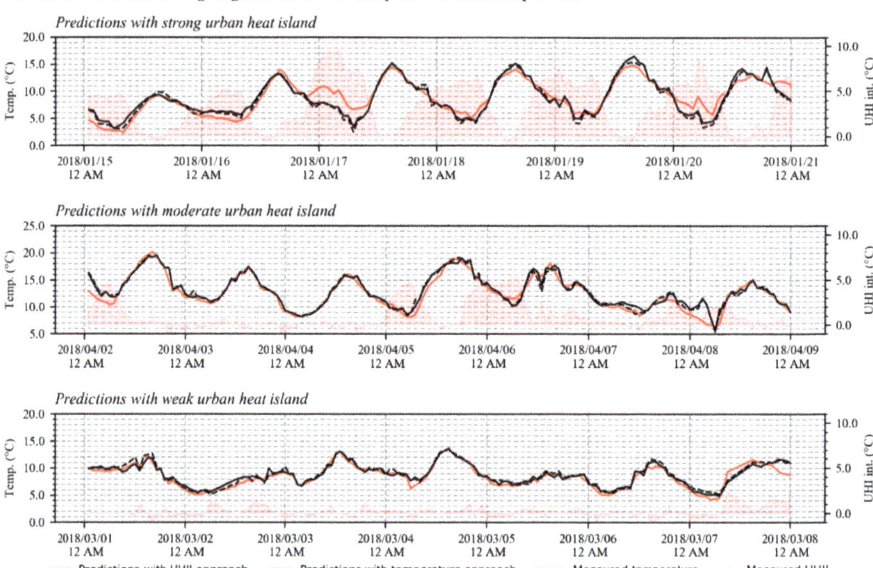

Figure 7. Results obtained with the models derived from the TEMP and the UHII approach for the site *Embajadores*. Three weeks were selected, each one representing a different atmospheric stability scenario. The timeframe used to train these models extends from July 2016 to September 2017.

Models targeting the UHI intensity got a slightly better score in the error metrics, with a reduction of the error between 6.4 and 11.7% (see Table 3). RMSE was 1.09 °C and 1.02 °C for the TEMP and the UHII approach, respectively. These results are in line with previous studies, such as in Kim and Baik (RMSE = 1.18 °C, [70]) or Demirezen et al. (RMSE = 1.29 °C, [75]), both modelling outdoor air temperature. The only exception is the coefficient of determination, which is extraordinarily high when targeting the temperature (R^2 = 0.99). This is also in line with previous studies (e.g., [75,77]) and it is further addressed in the discussion section.

Table 3. Metrics of the selected models targeting both the air temperature and the UHI intensity. Both models were trained using 12 months of data (July 2016–September 2017). The two variables regressed are modelled and monitored air temperatures.

Metrics		Model Targeting		Error Variation
		Temperature	UHI Intensity	
MAD	Median Absolute Deviation (°C)	0.60	0.53	−11.7%
MAE	Mean Absolute Error (°C)	0.81	0.74	−8.6%
RMSE	Root Mean Squared Error (°C)	1.09	1.02	−6.4%
R^2	Coefficient of Determination	0.99	0.79	+20.2%

Shortening the Training Dataset

The results presented above correspond to FNN models trained with one year of hourly data. So far, the TEMP and the UHII approach have proved to yield similar results. When training models with less data, however, differences started to arise. Results show that using 9 months instead of 12 months of data slightly increased the RMSE, with 0.9% and 2.4% for the TEMP and the UHII approach, respectively. When using 6 months of data the accuracy loss increased more markedly, especially in the case of the TEMP models (11.7% vs. 6.2%). The error kept growing exponentially when using 3 months of data, although the tendency was more accentuated and led to significant differences between both approaches (63.1% vs. 40.7%). A similar trend was observed for the MAE and MAD metrics, which can be found in Table 4.

Table 4. Relative accuracy loss when reducing the size of the training dataset for both the TEMP and the UHII approach. The number of months in each column establish the baseline of accuracy. The accuracy was obtained using the evaluation dataset.

	TEMP Approach				UHII Approach			
	RMSE				RMSE			
	12 months	9 months	6 months	3 months	12 months	9 months	6 months	3 months
12 months	0.0%				0.0%			
9 months	0.9%	0.0%			2.4%	0.0%		
6 months	11.7%	10.6%	0.0%		6.2%	3.8%	0.0%	
3 months	63.1%	61.6%	46.1%	0.0%	40.7%	37.5%	32.5%	0.0%
	MAE				MAE			
	12 months	9 months	6 months	3 months	12 months	9 months	6 months	3 months
12 months	0.0%				0.0%			
9 months	3.2%	0.0%			4.8%	0.0%		
6 months	14.0%	10.4%	0.0%		8.9%	4.0%	0.0%	
3 months	66.1%	60.9%	45.7%	0.0%	40.0%	33.7%	28.5%	0.0%
	MAD				MAD			
	12 months	9 months	6 months	3 months	12 months	9 months	6 months	3 months
12 months	0.0%				0.0%			
9 months	7.6%	0.0%			10.6%	0.0%		
6 months	17.7%	9.4%	0.0%		14.5%	3.5%	0.0%	
3 months	70.7%	58.7%	45.0%	0.0%	42.7%	29.0%	24.6%	0.0%

These results are the average error yielded by several models trained with shortened datasets and are relative to the accuracy of the models trained with 12 months of data. Figure 8 presents the models' accuracy absolute levels, including the accuracy of all models trained with each shortened dataset. As already noted, differences arise not only when reducing the training datasets, but also when changing from one approach to another. The large variability of error between the models trained with 3 months of data is noticeable, being more accentuated in the case of the TEMP approach. It seems that, depending on the data used during training, it is possible to obtain models with an acceptable overall accuracy (RMSE < 1.5 °C, in line with previously developed models) to others that it is not clear that they could be used to make a reasonable modelling (RMSE > 2 °C).

Yet, these results represent the average cumulative error over a year. A more detailed analysis of the accuracy of the models showed that their error is unevenly distributed over the months, losing accuracy outside the months for which they were trained. It was also observed that their results do not suffer excessively within the months for which they were trained, being comparable with models trained on more data. In that sense, Figure 9 shows the additional error yielded by models trained with only 3 months of data. For convenience purposes, these months were made coincident with the seasons of the year, and a model trained with all 12 months of data was used as a reference.

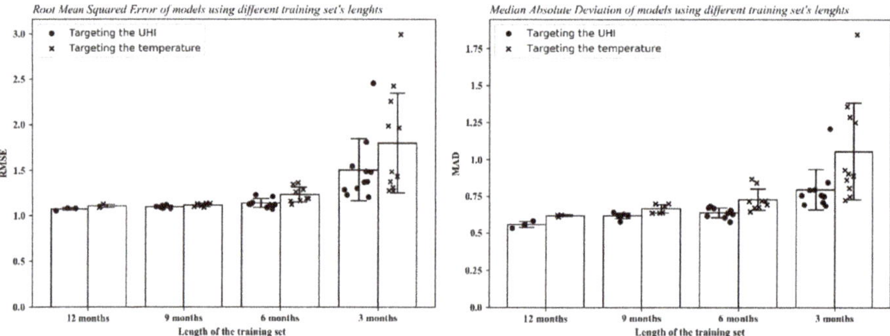

Figure 8. Comparison of the error obtained by several models, trained using different datasets of different length and differentiating between those targeting the air temperature and the UHI intensity. On the left is presented the RMSE. On the left is presented the RMSE. On the right is presented the MAD.

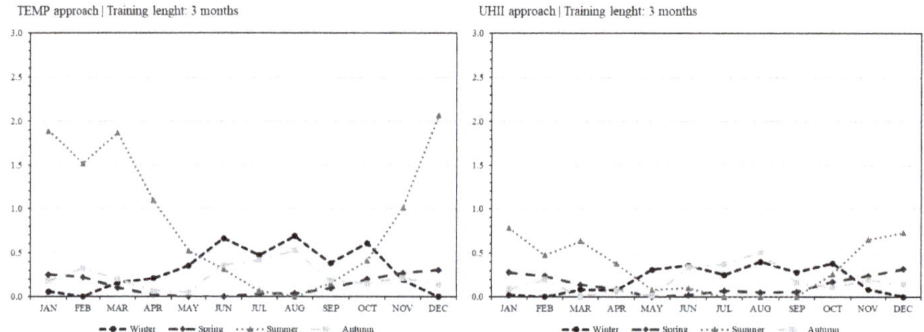

Figure 9. Additional error yielded by models trained with just 3 months of data, using both the TEMP and the UHII approach. These models are named according to the season they were trained on. The reference error baseline was established by the same ANN configuration trained with 12 months of data.

The results show that the models systematically tend to minimise their error within their season, with the RMSE gradually increasing as they move away from it. This is accentuated for models trained in winter and summer. The reason behind this could lie in the annual cyclical behaviour of temperatures: between solstices and equinoxes, temperatures remain at one extreme of the annual cycle, either at the high end of temperatures (summer) or at the low end (winter). Between the equinoxes and solstices (spring and autumn), though, the transition between the two extremes takes place. This could favour the training of the ANN, as it would extend pattern recognition to practically the entire annual temperature range, and where only the extremes would be at the expense of the neural network's ability to generalise and extrapolate its modelling capacity beyond what is known during its training.

This dynamic is noticeable in the case of the UHII approach as well, although it seems to be rather less pronounced. As it was pointed out in the introduction, Madrid's UHI does not seem to follow a seasonal pattern, which means it might reach its highest and lowest UHII intensities at any time during the year (see Figure A1). However, these UHI peaks depend on the meteorological conditions, thus the loss of accuracy registered by these UHII models seems to be likely related to the concentration of certain meteorological conditions during the training phase. In other words, these FNNs would have difficulties in refining the modelling if, within the three months of data used to train them, there is not a sufficiently large record of the different meteorological conditions that favour the occurrence of UHI.

The performance differences between the TEMP and the UHII approach are now clearly noticeable when plotting the data. In this respect, Figure 10 shows how the results of a TEMP model trained from May to August would produce quite precise results for June of the next year, like the ones obtained by models trained with 12 months of data. However, when trying to obtain the temperature profile in February, that same model barely captures the global trend. In that scenario, the UHII model, trained with the same three months of data, was able to fit to observed values with higher accuracy. It accumulated the error at the same moment as the models trained with 12 months of data, in many cases amplifying it. Despite the unusual distribution of temperatures and UHI intensities for that week, the UHII model was able to capture most of it, which turned to be surprising due to the relatively low amount of data used for its training.

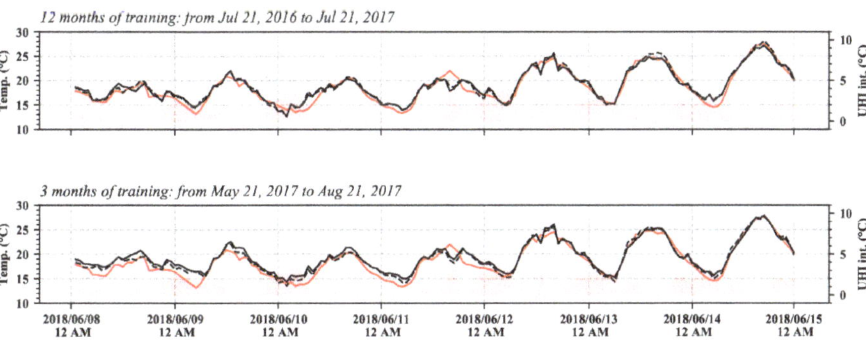

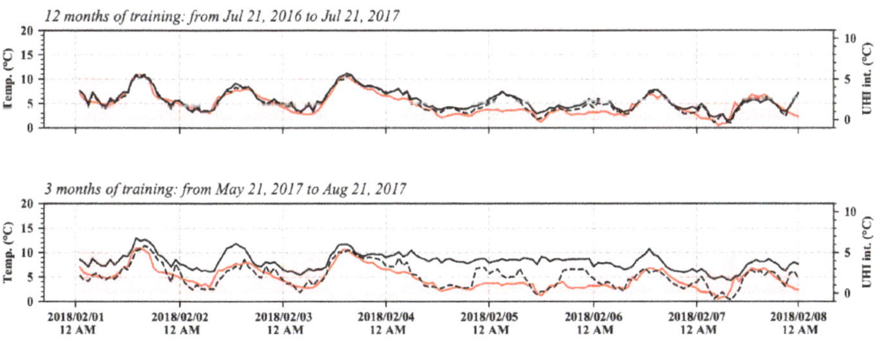

Figure 10. Comparison of TEMP and UHII models trained with 12 and 3 months of data for the site Embajadores. Results are presented for one week of June (top) and one week of February (bottom). The TEMP model trained on 3 months of data shows difficulties when modelling the temperatures further away from its training temporal range (i.e., February, bottom).

4. Discussion

The results of this research point towards the potential reduction of the training datasets without having a significant loss of accuracy. This could facilitate the work of urban climate researchers, thus promoting the development of shorter and simpler monitoring campaigns. This does not mean that it is preferable to use smaller amounts of data to train ANN models, but that their accuracy might not be compromised when they are trained in this manner. Although using large amounts of high-quality data is always desirable, in some cases it is not possible due to varying circumstances, such as budget constraints or human resources limitations. In this context, knowing where the accuracy limits of the models are when trained with fewer data might help researchers explore their experimental data or design new measurement campaigns in an efficient manner.

In this study we propose the use of empirical, FNN-based models to extend the temporal coverage of urban monitoring campaigns. These models, although limited for carrying out temporal predictions into the future, they can be used to adjust long-term records gathered outside the city to the urban context. This approach, the generation of long-term datasets by looking backwards, might be potentially useful in many disciplines, including the generation of site-specific weather files for building energy modelling [120–123], the downscaling of heat-related epidemiological studies to evaluate the effect of urban temperatures in health [4,124–127], or the identification and characterization of energy poor households in urban environments [128–132].

It is worth noting that the use of UHI intensity instead of outdoor temperature as the output of the FNN models yielded significantly better results mainly when reducing the size of the training dataset. The accuracy improvement was limited when using 9 or more months of data during the training phase. The benefits of targeting the UHI intensity with the FNN model are, therefore, linked to the potential of using smaller datasets to model outdoor urban temperatures. However, using the UHI intensity instead of the temperature as the output, sustained on the lower seasonality of the former, could be arguable. ANN are universal function approximators [133] and, for that reason, using one parameter or the other should not produce significant differences. Although this was mathematically demonstrated, Curry [134] showed that to model the seasonality of a time series with FNN would require a very large structure. This structure would grow exponentially when increasing the length of the dataset, since more turning points are likely to appear. In fact, Zhang and Qi [135] recommended not only to deseasonalize the time series, but also to remove its trend (if any). Nowadays, pre-processing the dataset to make it stationary before feeding the ANN is a very extended practice and has demonstrated to be very effective with RNN as well [88,136]. This approach might be helpful in the future for other studies such as Han et al. [86], where the UHI intensity could be used instead of the outdoor air temperature to remove much of the seasonality from their temperature forecasts. However, it is unclear whether they could be extended to FNNs that use a reference site for modelling outdoor urban temperatures without any time dependence. Other reasons, such as the range of temperatures or the concentration of meteorological stability of the training dataset, might explain the varying accuracy results between the TEMP and the UHII approach when training these types of models, especially when using just 3 months of training data.

In line with the latter, it seems that the selection of days with different meteorological conditions and at different times of the year might be more relevant for the modelling than the continuity of the monitoring campaign. Thus, it may be more appropriate that future studies work with shorter, discontinuous monitoring campaigns covering a wider range of meteorological situations rather than a single, continuous-over-time campaign that might concentrate in a specific time of the year. Results may also support the use of data from sources whose long-term continuity may be compromised (i.e., CWS). In these cases, it would be relevant not only to apply filtering techniques to reduce the risks of introducing outliers, but also to carry out frequency distribution analyses to ensure that all meteorological conditions are being included into the modelling.

Some attention should be drawn to the pertinence of using certain error metrics. Despite being very extended (e.g., [75–77]), the use of R^2 as a performance indicator could be misleading [137,138]. As it can be seen in Equation (1), R^2 relies both on the size of the residuals (SS_{res}, the actual deviation of the prediction from the observed values) and the total variance of the dependent variable (SS_{tot}):

$$R^2 = 1 - \frac{\sum(y_i - \hat{y}_i)^2}{\sum(y_i - \bar{y}_i)^2} = 1 - \frac{SS_{res}}{SS_{tot}} \tag{1}$$

Thus, obtaining a higher R^2 does not implicitly mean having less error (numerator), but might be the result of a higher variance of the output (denominator). This was observed in this study when comparing the two approaches. In the case of the TEMP approach, the variance of the output temperature (which ranges from -2 to 41 °C) is much higher that the variance of the UHI intensity (ranging from 0 to 8 °C). Furthermore, since the TEMP approach contains an input variable (airport temperature) that explains most of the variance of the output variable (urban temperature), the R^2 tends to be extremely high ($R^2 > 0.99$). This explains why significantly lower R^2 were obtained when using the UHII approach in spite of yielding better results with the rest of the performance indicators (RMSE, MAE, MAD).

Taking the above into consideration, it would be worthwhile to investigate whether the behaviour of the TEMP and the UHII models presented in this paper is only attributable to the case of Madrid or if, on the contrary, it might be common in other cities at different latitudes and climatic conditions. Some existing studies have identified strong seasonal differences in UHI intensities in other cities [139–142], while others have not found such differences [143–145]. In this respect, a strong annual seasonality of the UHI intensity would probably limit the capacity of the models to produce accurate results when trained with small datasets. On the contrary, in cities where air temperatures remain within a narrow range throughout the year, such as tropical regions, the TEMP approach might perform better.

5. Conclusions

Feed-forward neural networks were used in this study to model urban temperature time series from experimental data. The aim was to explore the reliability of these models in the context of low data availability, as well as the potential benefits from targeting the UHI intensity with these models. Results showed that, for the case study of Madrid, the training dataset could be reduced to 9 or even 6 months without compromising too much the accuracy of the FNN models, particularly when using the UHII approach (2.4% and 6.2% increase in RMSE, respectively).

Results showed that the UHII approach generally outperformed the TEMP approach. Overall, UHII models converged to lower error ratios with a smaller number of neurons, proving to be more effective at predicting the urban temperature of a reference site. When using the exact same configuration and structure, UHII models exhibited a significant increase in performance. TEMP models appeared to be quite seasonally dependent, thus facing more problems for modelling temperatures outside the training months. This was particularly relevant when trained on just 3 months of data, when the accuracy differences between UHII and TEMP models was at their highest. We argue that this could be related to the annual cyclical behaviour of temperatures. Targeting the UHI intensity with the FNNs instead, which in Madrid has shown to be almost stationary, seems to reduce uncertainty when modelling temperatures from a relatively small dataset.

The potential use of smaller datasets for training FNNs and still obtaining reliable results might benefit urban climate researchers since field measurements could be reduced in time and costs. Researchers might also take advantage of the accurate preliminary results that can be generated with relatively small datasets for speeding up their research, or for extending their measurements to other urban areas.

Author Contributions: Conceptualization, M.N.-P., A.M. and P.S.; methodology, M.N.-P., A.M. and P.S.; software, M.N.-P. and P.S.; validation, M.N.-P., A.M. and P.S.; formal analysis, M.N.-P., A.M. an P.S.; investigation, M.N.-P.; resources, M.N.-P. and C.S.-G.S.; data curation, M.N.-P.; writing—original draft preparation, M.N.-P., A.M., P.S. and C.S.-G.S.; writing—review and editing, M.N.-P., A.M., P.S., C.S.-G.S. and F.J.N.G.; visualization, M.N.-P.; supervision, A.M. and P.S.; project administration, M.N.-P., C.S.-G.S. and F.J.N.G.; funding acquisition, M.N.-P., C.S.-G.S. and F.J.N.G. All authors have read and agreed to the published version of the manuscript.

Funding: This research was funded by an FPU research grant (FPU15/05052) and by a research visit grant (EST17/00825), both from the Spanish Ministry of Education, Culture and Sport. This research was also supported by the MODIFICA research project (BIA2013-41732-R), funded by the Spanish Ministry of Economy and Competitiveness.

Institutional Review Board Statement: Not applicable.

Informed Consent Statement: Not applicable.

Acknowledgments: The authors would like to thank the Spanish National Meteorological Agency (AEMET) for providing access to their meteorological data, and Luis Tejero Encinas and Juan Azcárate Luxán, from the Madrid City Council' Subdivision of Energy and Climate Change, for their support with the urban measurements.

Conflicts of Interest: The authors declare no conflict of interest.

Appendix A

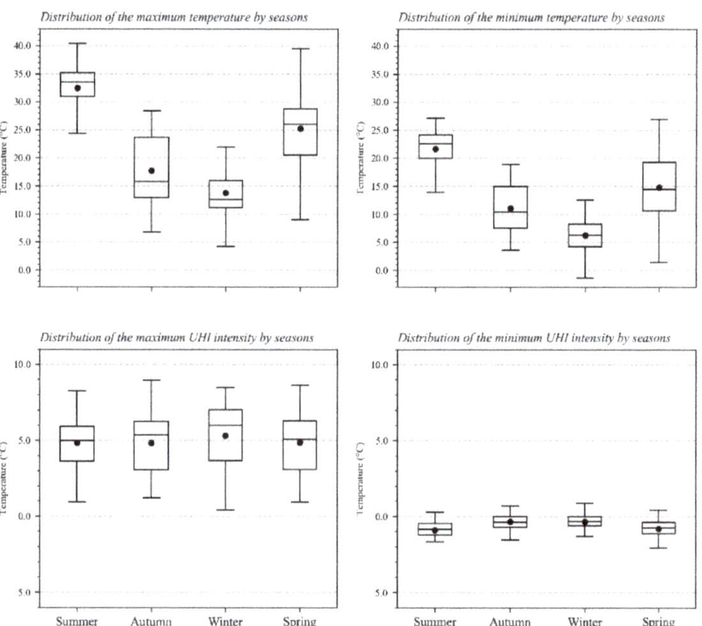

Figure A1. Seasonal fluctuations of the daily maximum and minimum temperature (top) and UHI intensity (bottom) between August 2016 and July 2018.

References

1. Reckien, D.; Salvia, M.; Heidrich, O.; Church, J.M.; Pietrapertosa, F.; De Gregorio-Hurtado, S.; D'Alonzo, V.; Foley, A.; Simoes, S.G.; Krkoška Lorencová, E.; et al. How are cities planning to respond to climate change? Assessment of local climate plans from 885 cities in the EU-28. *J. Clean. Prod.* **2018**, *191*, 207–219. [CrossRef]
2. Santamouris, M.; Cartalis, C.; Synnefa, A. Local urban warming, possible impacts and a resilience plan to climate change for the historical center of Athens, Greece. *Sustain. Cities Soc.* **2015**, *19*, 281–291. [CrossRef]
3. Moran, D.; Kanemoto, K.; Jiborn, M.; Wood, R.; Többen, J.; Seto, K.C. Carbon footprints of 13,000 cities. *Environ. Res. Lett.* **2018**, *13*. [CrossRef]

4. Macintyre, H.; Heaviside, C.; Cai, X.; Phalkey, R. Comparing temperature-related mortality impacts of cool roofs in winter and summer in a highly urbanized European region for present and future climate. *Environ. Int.* **2021**, *154*, 106606. [CrossRef]

5. Sánchez-Guevara, C.; Núñez Peiró, M.; Taylor, J.; Mavrogianni, A.; Neila González, J. Assessing population vulnerability towards summer energy poverty: Case studies of Madrid and London. *Energy Build.* **2019**, *190*, 132–143. [CrossRef]

6. Hsu, A.; Sheriff, G.; Chakraborty, T.; Manya, D. Disproportionate exposure to urban heat island intensity across major US cities. *Nat. Commun.* **2021**, *12*, 2721. [CrossRef]

7. Grimm, N.B.; Faeth, S.H.; Golubiewski, N.E.; Redman, C.L.; Wu, J.; Bai, X.; Briggs, J.M. Global change and the ecology of cities. *Science* **2008**, *319*, 756–760. [CrossRef] [PubMed]

8. Youngsteadt, E.; Dale, A.G.; Terando, A.J.; Dunn, R.R.; Frank, S.D. Do cities simulate climate change? A comparison of herbivore response to urban and global warming. *Glob. Chang. Biol.* **2015**, *21*, 97–105. [CrossRef]

9. Blocken, B. Computational Fluid Dynamics for urban physics: Importance, scales, possibilities, limitations and ten tips and tricks towards accurate and reliable simulations. *Build. Environ.* **2015**, *91*, 219–245. [CrossRef]

10. Doan, V.Q.; Kusaka, H.; Nguyen, T.M. Roles of past, present, and future land use and anthropogenic heat release changes on urban heat island effects in Hanoi, Vietnam: Numerical experiments with a regional climate model. *Sustain. Cities Soc.* **2019**, *47*, 101479. [CrossRef]

11. Ampatzidis, P.; Kershaw, T. A review of the impact of blue space on the urban microclimate. *Sci. Total Environ.* **2020**, *730*, 139068. [CrossRef]

12. Boehme, P.; Berger, M.; Massier, T. Estimating the building based energy consumption as an anthropogenic contribution to urban heat islands. *Sustain. Cities Soc.* **2015**, *19*, 373–384. [CrossRef]

13. Best, M.J.; Grimmmmond, C.S.B. Key conclusions of the first international urban land surface model comparison project. *Bull. Am. Meteorol. Soc.* **2015**, *96*, 805–819. [CrossRef]

14. Jandaghian, Z.; Berardi, U. Comparing urban canopy models for microclimate simulations in Weather Research and Forecasting Models. *Sustain. Cities Soc.* **2020**, *55*, 102025. [CrossRef]

15. Chen, Y.; Zheng, B.; Hu, Y. Numerical simulation of Local Climate Zone cooling achieved through modification of trees, albedo and green roofs-a case study of Changsha, China. *Sustainability* **2020**, *12*, 2752. [CrossRef]

16. Tsoka, S.; Tsikaloudaki, A.; Theodosiou, T. Analyzing the ENVI-met microclimate model's performance and assessing cool materials and urban vegetation applications—A review. *Sustain. Cities Soc.* **2018**, *43*, 55–76. [CrossRef]

17. Mirzaei, P.A. Recent challenges in modeling of urban heat island. *Sustain. Cities Soc.* **2015**, *19*, 200–206. [CrossRef]

18. Toparlar, Y.; Blocken, B.; Maiheu, B.; van Heijst, G.J.F. A review on the CFD analysis of urban microclimate. *Renew. Sustain. Energy Rev.* **2017**, *80*, 1613–1640. [CrossRef]

19. Lauzet, N.; Rodler, A.; Musy, M.; Azam, M.H.; Guernouti, S.; Mauree, D.; Colinart, T. How building energy models take the local climate into account in an urban context—A review. *Renew. Sustain. Energy Rev.* **2019**, *116*, 109390. [CrossRef]

20. Mirzaei, P.A.; Haghighat, F. Approaches to study Urban Heat Island—Abilities and limitations. *Build. Environ.* **2010**, *45*, 2192–2201. [CrossRef]

21. Stewart, I.D. Why should urban heat island researchers study history? *Urban Clim.* **2019**, *30*, 100484. [CrossRef]

22. Velasco, E. Go to field, look around, measure and then run models. *Urban Clim.* **2018**, *24*, 231–236. [CrossRef]

23. Muller, C.L.; Chapman, L.; Grimmond, C.S.B.; Young, D.T.; Cai, X. Sensors and the city: A review of urban meteorological networks. *Int. J. Climatol.* **2013**, *33*, 1585–1600. [CrossRef]

24. Gobakis, K.; Kolokotsa, D.; Synnefa, A.; Saliari, M.; Giannopoulou, K.; Santamouris, M. Development of a model for urban heat island prediction using neural network techniques. *Sustain. Cities Soc.* **2011**, *1*, 104–115. [CrossRef]

25. Giridharan, R.; Kolokotroni, M. Urban heat island characteristics in London during winter. *Sol. Energy* **2009**, *83*, 1668–1682. [CrossRef]

26. Kolokotroni, M.; Giridharan, R. Urban heat island intensity in London: An investigation of the impact of physical characteristics on changes in outdoor air temperature during summer. *Sol. Energy* **2008**, *82*, 986–998. [CrossRef]

27. Zhou, X.; Okaze, T.; Ren, C.; Cai, M.; Ishida, Y.; Watanabe, H.; Mochida, A. Evaluation of urban heat islands using local climate zones and the influence of sea-land breeze. *Sustain. Cities Soc.* **2020**, *55*, 102060. [CrossRef]

28. Skarbit, N.; Stewart, I.D.; Unger, J.; Gál, T. Employing an urban meteorological network to monitor air temperature conditions in the "local climate zones" of Szeged, Hungary. *Int. J. Climatol.* **2017**, *37*, 582–596. [CrossRef]

29. Chen, G.; He, M.; Li, N.; He, H.; Cai, Y.; Zheng, S. A method for selecting the typical days with full urban heat island development in hot and humid area, case study in Guangzhou, China. *Sustainability* **2021**, *13*, 320. [CrossRef]

30. Chao, C.C.; Hung, K.A.; Chen, S.Y.; Lin, F.Y.; Lin, T.P. Application of a high-density temperature measurement system for the management of the kaohsiung house project. *Sustainability* **2021**, *13*, 960. [CrossRef]

31. Borbora, J.; Das, A.K. Summertime Urban Heat Island study for Guwahati City, India. *Sustain. Cities Soc.* **2014**, *11*, 61–66. [CrossRef]

32. Beck, C.; Straub, A.; Breitner, S.; Cyrys, J.; Philipp, A.; Rathmann, J.; Schneider, A.; Wolf, K.; Jacobeit, J. Air temperature characteristics of local climate zones in the Augsburg urban area (Bavaria, southern Germany) under varying synoptic conditions. *Urban Clim.* **2018**, *25*, 152–166. [CrossRef]

33. Yang, X.; Yao, L.; Jin, T.; Peng, L.L.H.; Jiang, Z.; Hu, Z.; Ye, Y. Assessing the thermal behavior of different local climate zones in the Nanjing metropolis, China. *Build. Environ.* **2018**, *137*, 171–184. [CrossRef]

34. Yang, X.; Peng, L.L.H.; Chen, Y.; Yao, L.; Wang, Q. Air humidity characteristics of local climate zones: A three-year observational study in Nanjing. *Build. Environ.* **2020**, *171*, 106661. [CrossRef]
35. van der Heijden, M.G.M.; Blocken, B.; Hensen, J.L.M. Towards the integration of the urban heat island in building energy simulations. In Proceedings of the Building Simulation 2013: 13th Conference of the International Building Performance Simulation Association IBPSA, Chamberry, Ference, 25–28 August 2013; pp. 1006–1013.
36. Fenner, D.; Meier, F.; Scherer, D.; Polze, A. Spatial and temporal air temperature variability in Berlin, Germany, during the years 2001–2010. *Urban Clim.* **2014**, *10*, 308–331. [CrossRef]
37. Meier, F.; Fenner, D.; Grassmann, T.; Otto, M.; Scherer, D. Crowdsourcing air temperature from citizen weather stations for urban climate research. *Urban Clim.* **2017**, *19*, 170–191. [CrossRef]
38. Muller, C.L.; Chapman, L.; Johnston, S.; Kidd, C.; Illingworth, S.; Foody, G.; Overeem, A.; Leigh, R.R. Crowdsourcing for climate and atmospheric sciences: Current status and future potential. *Int. J. Climatol.* **2015**, *35*, 3185–3203. [CrossRef]
39. Fenner, D.; Meier, F.; Bechtel, B.; Otto, M.; Scherer, D. Intra and inter "local climate zone" variability of air temperature as observed by crowdsourced citizen weather stations in Berlin, Germany. *Meteorol. Z.* **2017**, *26*, 525–547. [CrossRef]
40. Nipen, T.N.; Seierstad, I.A.; Lussana, C.; Kristiansen, J.; Hov, Ø. Adopting citizen observations in operational weather prediction. *Bull. Am. Meteorol. Soc.* **2020**, *101*, E43–E57. [CrossRef]
41. Bell, S.; Cornford, D.; Bastin, L. How good are citizen weather stations? Addressing a biased opinion. *Weather* **2015**, *70*, 75–84. [CrossRef]
42. Chapman, L.; Bell, C.; Bell, S. Can the crowdsourcing data paradigm take atmospheric science to a new level? A case study of the urban heat island of London quantified using Netatmo weather stations. *Int. J. Climatol.* **2017**, *37*, 3597–3605. [CrossRef]
43. Kousis, I.; Pigliautile, I.; Pisello, A.L. Intra-urban microclimate investigation in urban heat island through a novel mobile monitoring system. *Nat. Sci. Rep.* **2021**, *11*, 9732. [CrossRef]
44. Yadav, N.; Sharma, C. Spatial variations of intra-city urban heat island in megacity Delhi. *Sustain. Cities Soc.* **2018**, *37*, 298–306. [CrossRef]
45. Schneider, R.; Taylor, J.; Davies, M.; Mavrogianni, A.; Milner, J.; Dos Santos, R.S.; Taylor, J.; Davies, M.; Mavrogianni, A.; Milner, J. The variation of air and surface temperatures in London within a 1 km grid using vehicle-transect and ASTER data. In *Proceedings of the 2017 Joint Urban Remote Sensing Event, JURSE 2017, Dubai, United Arab Emirates, 6–8 March 2017*; Institute of Electrical and Electronics Engineers Inc.: Manhattan, NY, USA, 2017; pp. 6–9.
46. Romero Rodríguez, L.; Sánchez Ramos, J.; Sánchez de la Flor, F.J.; Álvarez Domínguez, S. Analyzing the urban heat Island: Comprehensive methodology for data gathering and optimal design of mobile transects. *Sustain. Cities Soc.* **2020**, *55*, 102027. [CrossRef]
47. Heusinkveld, B.G.; Van Hove, L.W.A.; Jacobs, C.M.J.; Steeneveld, G.J.; El-Bers, J.A.; Moors, E.J.; Holtslag, A.A.M. Use of a mobile platform for assessing urban heat stress in Rotterdam. In Proceedings of the 7th Conference on Biometeorology, Freiburg, Germany, 12–14 April 2010; Volume 12, pp. 433–438.
48. Chow, W.T.L.; Pope, R.L.; Martin, C.A.; Brazel, A.J. Observing and modeling the nocturnal park cool island of an arid city: Horizontal and vertical impacts. *Theor. Appl. Climatol.* **2011**, *103*, 197–211. [CrossRef]
49. Brandsma, T.; Wolters, D. Measurement and statistical modeling of the urban heat island of the city of Utrecht (The Netherlands). *J. Appl. Meteorol. Climatol.* **2012**, *51*, 1046–1060. [CrossRef]
50. Fabbri, K.; Costanzo, V. Drone-assisted infrared thermography for calibration of outdoor microclimate simulation models. *Sustain. Cities Soc.* **2020**, *52*, 101855. [CrossRef]
51. Zhou, D.; Xiao, J.; Bonafoni, S.; Berger, C.; Deilami, K.; Zhou, Y.; Frolking, S.; Yao, R.; Qiao, Z.; Sobrino, J.A. Satellite remote sensing of surface urban heat islands: Progress, challenges, and perspectives. *Remote Sens.* **2019**, *11*, 48. [CrossRef]
52. László, E.; Szegedi, S. A multivariate linear regression model of mean maximum urban heat island: A case study of Beregszász (Berehove), Ukraine. *Idojaras* **2015**, *119*, 409–423.
53. Levermore, G.; Parkinson, J. The urban heat island of London, an empirical model. *Build. Serv. Eng. Res. Technol.* **2019**, *40*, 290–295. [CrossRef]
54. Levermore, G.J.; Parkinson, J.B. An empirical model for the urban heat island intensity for a site in Manchester. *Build. Serv. Eng. Res. Technol.* **2017**, *38*, 21–31. [CrossRef]
55. Romero Rodríguez, L.; Sánchez Ramos, J.; Molina Félix, J.L.; Álvarez Domínguez, S. Urban-scale air temperature estimation: Development of an empirical model based on mobile transects. *Sustain. Cities Soc.* **2020**, *63*, 102471. [CrossRef]
56. Bernard, J.; Musy, M.; Calmet, I.; Bocher, E.; Keravec, P. Urban heat island temporal and spatial variations: Empirical modeling from geographical and meteorological data. *Build. Environ.* **2017**, *125*, 423–438. [CrossRef]
57. Chun, B.; Guldmann, J.M. Spatial statistical analysis and simulation of the urban heat island in high-density central cities. *Landsc. Urban Plan.* **2014**, *125*, 76–88. [CrossRef]
58. Gardes, T.; Schoetter, R.; Hidalgo, J.; Long, N.; Marquès, E.; Masson, V. Statistical prediction of the nocturnal urban heat island intensity based on urban morphology and geographical factors—An investigation based on numerical model results for a large ensemble of French cities. *Sci. Total Environ.* **2020**, *737*, 139253. [CrossRef] [PubMed]
59. Jin, H.; Cui, P.; Wong, N.H.; Ignatius, M. Assessing the effects of urban morphology parameters on microclimate in Singapore to control the urban heat island effect. *Sustainability* **2018**, *10*, 206. [CrossRef]

60. Straub, A.; Berger, K.; Breitner, S.; Cyrys, J.; Geruschkat, U.; Jacobeit, J.; Kühlbach, B.; Kusch, T.; Philipp, A.; Schneider, A.; et al. Statistical modelling of spatial patterns of the urban heat island intensity in the urban environment of Augsburg, Germany. *Urban Clim.* **2019**, *29*, 100491. [CrossRef]
61. Chang, J.M.-H.; Lam, Y.F.; Lau, S.P.-W.; Wong, W.-K. Development of fine-scale spatiotemporal temperature forecast model with urban climatology and geomorphometry in Hong Kong. *Urban Clim.* **2021**, *37*, 100816. [CrossRef]
62. Ho, H.C.; Knudby, A.; Sirovyak, P.; Xu, Y.; Hodul, M.; Henderson, S.B. Mapping maximum urban air temperature on hot summer days. *Remote Sens. Environ.* **2014**, *154*, 38–45. [CrossRef]
63. Lai, J.; Zhan, W.; Quan, J.; Bechtel, B.; Wang, K.; Zhou, J.; Huang, F.; Chakraborty, T.; Liu, Z.; Lee, X. Statistical estimation of next-day nighttime surface urban heat islands. *ISPRS J. Photogramm. Remote Sens.* **2021**, *176*, 182–195. [CrossRef]
64. Zhou, J.; Zhou, J.; Chen, Y.; Wang, J.; Zhan, W.; Wang, J. Maximum Nighttime Urban Heat Island (UHI) Intensity Simulation by Integrating Remotely Sensed Data and Meteorological Observations. *IEEE J. Sel. Top. Appl. Earth Obs. Remote Sens.* **2011**, *4*, 138–146. [CrossRef]
65. Chen, Z.; Zhu, Z.; Jiang, H.; Sun, S. Estimating daily reference evapotranspiration based on limited meteorological data using deep learning and classical machine learning methods. *J. Hydrol.* **2020**, *591*, 125286. [CrossRef]
66. Venter, Z.S.; Brousse, O.; Esau, I.; Meier, F. Hyperlocal mapping of urban air temperature using remote sensing and crowdsourced weather data. *Remote Sens. Environ.* **2020**, *242*, 111791. [CrossRef]
67. Mihalakakou, G.; Santamouris, M.; Asimakopoulos, D. Modeling ambient air temperature time series using neural networks. *J. Geophys. Res.* **1998**, *103*, 19509–19517. [CrossRef]
68. Mihalakakou, G.; Flocas, H.A.; Santamouris, M.; Helmis, C.G. Application of Neural Networks to the Simulation of the Heat Island over Athens, Greece, Using Synoptic Types as a Predictor. *J. Appl. Meteorol.* **2002**, *41*, 519–527. [CrossRef]
69. Santamouris, M.; Mihalakakou, G.; Papanikolaou, N.; Asimakopoulos, D.N. A neural network approach for modeling the Heat Island phenomenon in urban areas during the summer period. *Geophys. Res. Lett.* **1999**, *26*, 337. [CrossRef]
70. Kim, Y.-H.; Baik, J.-J. Maximum Urban Heat Island Intensity in Seoul. *J. Appl. Meteorol.* **2002**, *41*, 651–659. [CrossRef]
71. Kolokotroni, M.; Davies, M.; Croxford, B.; Bhuiyan, S.; Mavrogianni, A. A validated methodology for the prediction of heating and cooling energy demand for buildings within the Urban Heat Island: Case-study of London. *Sol. Energy* **2010**, *84*, 2246–2255. [CrossRef]
72. Kolokotroni, M.; Zhang, Y.; Giridharan, R. Heating and cooling degree day prediction within the London urban heat island area. *Build. Serv. Eng. Res. Technol.* **2009**, *30*, 183–202. [CrossRef]
73. Kolokotroni, M.; Zhang, Y.; Watkins, R. The London Heat Island and building cooling design. *Sol. Energy* **2007**, *81*, 102–110. [CrossRef]
74. Demirezen, G.; Fung, A.S. Application of artificial neural network in the prediction of ambient temperature for a cloud-based smart dual fuel switching system. *Energy Procedia* **2019**, *158*, 3070–3075. [CrossRef]
75. Demirezen, G.; Fung, A.S.; Deprez, M. Development and optimization of artificial neural network algorithms for the prediction of building specific local temperature for HVAC control. *Int. J. Energy Res.* **2020**, *44*, 8513–8531. [CrossRef]
76. Papantoniou, S.; Kolokotsa, D. Prediction of outdoor air temperature using neural networks: Application in 4 European cities. *Energy Build.* **2016**, *114*, 72–79. [CrossRef]
77. Erdemir, D.; Ayata, T. Prediction of temperature decreasing on a green roof by using artificial neural network. *Appl. Therm. Eng.* **2017**, *112*, 1317–1325. [CrossRef]
78. Mihalakakou, G.; Santamouris, M.; Papanikolaou, N.; Cartalis, C.; Tsangrassoulis, A. Simulation of the Urban Heat Island Phenomenon in Mediterranean Climates. *Pure Appl. Geophys.* **2004**, *161*, 429–451. [CrossRef]
79. Jang, J.; Viau, A.A.; Anctil, F. Neural network estimation of air temperatures from AVHRR data. *Int. J. Remote Sens.* **2004**, *25*, 4541–4554. [CrossRef]
80. Zhao, D. Analysis of thermal environment and urban heat island using remotely sensed imagery over the north and south slope of the Qinling Mountain, China. In Proceedings of the 2007 IEEE International Geoscience and Remote Sensing Symposium, Barcelona, Spain, 23–27 July 2007; pp. 655–658.
81. Beccali, G.; Cellura, M.; Culotta, S.; Brano, V.L.; Marvuglia, A. A Web-Based Autonomous Weather Monitoring System of the Town of Palermo and Its Utilization for Temperature Nowcasting. In *Computational Science and Its Applications—ICCSA 2008*; Gervasi, O., Murgante, B., Laganà, A., Taniar, D., Mun, Y., Gavrilova, M.L., Eds.; Springer: Berlin, Germany, 2008; pp. 65–80.
82. Cellura, M.; Culotta, S.; Lo Brano, V.; Marvuglia, A.; Energetiche, R. Nonlinear Black-Box Models for Short-Term Forecasting of Air Temperature in the Town of Palermo. In *Geocomputation, Sustainability and Environmental Planning*; Murgante, B., Borruso, G., Lapucci, A., Eds.; Springer: Berlin, Germany, 2011; pp. 183–204.
83. Shao, B.; Zhang, M.; Mi, Q.; Xiang, N. Prediction and Visualization for Urban Heat Island. In *Transactions on Edutainment VI. Lecture Notes in Computer Science, Volume 6758*; Springer: Berlin, Germany, 2011; pp. 1–11.
84. Lee, Y.Y.; Kim, J.T.; Yun, G.Y. The neural network predictive model for heat island intensity in Seoul. *Energy Build.* **2016**, *110*, 353–361. [CrossRef]
85. Schuch, F.; Marpu, P.; Masri, D.; Afshari, A. Estimation of Urban Air Temperature from a Rural Station Using Remotely Sensed Thermal Infrared Data. *Energy Procedia* **2017**, *143*, 519–525. [CrossRef]
86. Han, J.M.; Ang, Y.Q.; Malkawi, A.; Samuelson, H.W. Using recurrent neural networks for localized weather prediction with combined use of public airport data and on-site measurements. *Build. Environ.* **2021**, *192*, 107601. [CrossRef]

87. ISO Online Browsing Platform. Available online: https://www.iso.org/obp/ui/#search (accessed on 14 January 2021).

88. Hewamalage, H.; Bergmeir, C.; Bandara, K. Recurrent Neural Networks for Time Series Forecasting: Current status and future directions. *Int. J. Forecast.* **2021**, *37*, 388–427. [CrossRef]

89. López Gómez, A.; López Gómez, J.; Fernández García, F.; Arroyo Ilera, F. *El Clima Urbano de Madrid: La Isla de Calor*; CSIC: Madrid, Spain, 1988; ISBN 978-84-00-07521-7.

90. Yagüe, C.; Zurita, E.; Martinez, A. Statistical analysis of the Madrid urban heat island. *Atmos. Environ.* **1991**, *25*, 327–332. [CrossRef]

91. Fernández García, F.; Montálvez, J.P.; González-Rouco, F.J.; Valero, F. A PCA Analysis of the UHI Form of Madrid. In Proceedings of the 5th International Conference on Urban Climate, Lodz, Poland, 1–5 September 2003; pp. 1–4.

92. Núñez Peiró, M.; Sánchez-Guevara Sánchez, C.; Neila González, F.J. *Update of the Urban Heat Island of Madrid and Its Influence on the Building's Energy Simulation*; Springer: New York, NY, USA, 2017; ISBN 9783319514420.

93. López Gómez, A.; López Gómez, J.; Fernández García, F.; Moreno Jiménez, A. *El Clima Urbano: Teledetección de la isla de Calor en Madrid*; MOPT: Madrid, Spain, 1993; ISBN 978-84-00-07521-7.

94. Sobrino, J.A.; Oltra-Carrió, R.; Sòria, G.; Jiménez-Muñoz, J.C.; Franch, B.; Hidalgo, V.; Mattar, C.; Julien, Y.; Cuenca, J.; Romaguera, M.; et al. Evaluation of the surface urban heat island effect in the city of Madrid by thermal remote sensing. *Int. J. Remote Sens.* **2013**, *34*, 3177–3192. [CrossRef]

95. Salamanca, F.; Martilli, A. A new Building Energy Model coupled with an Urban Canopy Parameterization for urban climate simulations-part II. Validation with one dimension off-line simulations. *Theor. Appl. Climatol.* **2010**, *99*, 345–356. [CrossRef]

96. Salamanca, F.; Martilli, A.; Yagüe, C. A numerical study of the Urban Heat Island over Madrid during the DESIREX (2008) campaign with WRF and an evaluation of simple mitigation strategies. *Int. J. Climatol.* **2011**, *32*, 2372–2386. [CrossRef]

97. Núñez-Peiró, M.; Sánchez-Guevara Sánchez, C.; Neila González, F.J. Hourly evolution of intra-urban temperature variability across the local climate zones. The case of Madrid. *Urban Clim.* **2021**, *39*, 100921. [CrossRef]

98. Stewart, I.D.; Oke, T.R. Local climate zones for urban temperature studies. *Bull. Am. Meteorol. Soc.* **2012**, *93*, 1879–1900. [CrossRef]

99. Oke, T.R. *Initial Guidance to Obtain Representative Meteorological Observations at Urban Sites (WMO/TD No. 1250)*; WMO: Geneva, Switzerland, 2006.

100. WMO. *Guide to Meteorological Instruments and Methods of Observation (WMO No. 8)*; WMO: Geneva, Switzerland, 2017. Available online: http://www.posmet.ufv.br/wp-content/uploads/2016/09/MET-474-WMO-Guide.pdf (accessed on 15 July 2021).

101. Núñez Peiró, M.; Sánchez-Guevara Sánchez, C.; Neila González, F.J. Source area definition for local climate zones studies. A systematic review. *Build. Environ.* **2019**, *148*, 258–285. [CrossRef]

102. Alexander, P.J.; Bechtel, B.; Chow, W.T.L.; Fealy, R.; Mills, G. Linking urban climate classification with an urban energy and water budget model: Multi-site and multi-seasonal evaluation. *Urban Clim.* **2016**, *17*, 196–215. [CrossRef]

103. Kaplan, S.; Peeters, A.; Erell, E. Predicting air temperature simultaneously for multiple locations in an urban environment: A bottom up approach. *Appl. Geogr.* **2016**, *76*, 62–74. [CrossRef]

104. WMO. *Guide to the Global Observing System (WMO No. 488)*; WMO: Geneva, Switzerland, 2017. Available online: https://library.wmo.int/doc_num.php?explnum_id=4236 (accessed on 15 July 2021).

105. Aguilar, E.; Auer, I.; Brunet, M.; Peterson, T.C.; Wieringa, J. *Guidelines on Climate Metadata and Homogenization (WMO/TD No. 1186)*; WMO: Geneva, Switzerland, 2003. Available online: https://library.wmo.int/doc_num.php?explnum_id=9252 (accessed on 15 July 2021).

106. WMO. *General Meteorological Standards and Recommended Practices (WMO No. 49)*; WMO: Geneva, Switzerland, 2018; Volume I. Available online: https://library.wmo.int/doc_num.php?explnum_id=10113 (accessed on 15 July 2021).

107. WMO. *WIGOS Metadata Standard 2019*; WMO: Geneva, Switzerland, 2019. Available online: https://library.wmo.int/doc_num.php?explnum_id=10109 (accessed on 15 July 2021).

108. Brousse, O.; Martilli, A.; Foley, M.; Mills, G.; Bechtel, B. WUDAPT, an efficient land use producing data tool for mesoscale models? Integration of urban LCZ in WRF over Madrid. *Urban Clim.* **2016**, *17*, 116–134. [CrossRef]

109. Oke, T.R.; Mills, G.; Christen, A.; Voogt, J.A. *Urban Climates*; Cambridge University Press: Cambridge, UK, 2017; ISBN 9781139016476.

110. Sundborg, A. Local Climatological Studies of the Temperature Conditions in an Urban Area. *Tellus* **1950**, *2*, 222–232. [CrossRef]

111. Zhou, M.; Qu, X.; Li, X. A recurrent neural network based microscopic car following model to predict traffic oscillation. *Transp. Res. Part C Emerg. Technol.* **2017**, *84*, 245–264. [CrossRef]

112. Shi, Y.; Ren, C.; Lau, K.K.L.; Ng, E. Investigating the influence of urban land use and landscape pattern on PM2.5 spatial variation using mobile monitoring and WUDAPT. *Landsc. Urban Plan.* **2019**, *189*, 15–26. [CrossRef]

113. Afshari, A.; Ramirez, N. Improving the accuracy of simplified urban canopy models for arid regions using site-specific prior information. *Urban Clim.* **2021**, *35*, 100722. [CrossRef]

114. Olden, J.D.; Jackson, D.A. Illuminating the "black box": Understanding variable contributions in artificial neural networks. *Ecol. Modell.* **2002**, *154*, 135–150. [CrossRef]

115. Shanker, M.S.; Hu, M.Y.; Hung, M.S. Effect of data standardization on neural network training. *Omega* **1996**, *24*, 385–397. [CrossRef]

116. Abadi, M.; Barham, P.; Chen, J.; Chen, Z.; Davis, A.; Dean, J.; Devin, M.; Ghemawat, S.; Irving, G.; Isard, M.; et al. TensorFlow: A System for Large-Scale Machine Learning. In Proceedings of the 12th USENIX Symposium on Operating Systems Design and Implementation (OSDI'16), Savannah, GA, USA, 2–4 November 2016.
117. Chollet, F. Keras. Available online: https://keras.io (accessed on 28 February 2021).
118. Cortez, P.; Embrechts, M.J. Using sensitivity analysis and visualization techniques to open black box data mining models. *Inf. Sci.* **2013**, *225*, 1–17. [CrossRef]
119. Gerald, W.; Davis, J. Sensitivity Analysis in Neural Net Solutions. *IEEE Trans. Syst. Man. Cybern.* **1989**, *19*, 1078–1082. [CrossRef]
120. Ferrando, M.; Causone, F.; Hong, T.; Chen, Y. Urban building energy modeling (UBEM) tools: A state-of-the-art review of bottom-up physics-based approaches. *Sustain. Cities Soc.* **2020**, *62*, 102408. [CrossRef]
121. Erba, S.; Causone, F.; Armani, R. The effect of weather datasets on building energy simulation outputs. *Energy Procedia* **2017**, *134*, 545–554. [CrossRef]
122. Taylor, J.; Davies, M.; Mavrogianni, A.; Chalabi, Z.; Biddulph, P.; Oikonomou, E.; Das, P.; Jones, B. The relative importance of input weather data for indoor overheating risk assessment in dwellings. *Build. Environ.* **2014**, *76*, 81–91. [CrossRef]
123. Bienvenido-Huertas, D.; Marín-García, D.; Carretero-Ayuso, M.J.; Rodríguez-Jiménez, C.E. Climate classification for new and restored buildings in Andalusia: Analysing the current regulation and a new approach based on k-means. *J. Build. Eng.* **2021**, *43*, 102829. [CrossRef]
124. López-Bueno, J.A.; Linares, C.; Sánchez-Guevara, C.; Sánchez-Martínez, G.; Mirón, I.J.; Núñez Peiró, M.; Valero, I.; Díaz, J. The effect of cold waves on daily mortality in districts in Madrid considering sociodemographic variables. *Sci. Total Environ.* **2020**, *749*, 142364. [CrossRef] [PubMed]
125. López-Bueno, J.A.; Díaz, J.; Sánchez-Guevara, C.; Sánchez-Martínez, G.; Franco, M.; Gullón, P.; Núñez Peiró, M.; Valero, I.; Linares, C. The impact of heat waves on daily mortality in districts in Madrid: The effect of sociodemographic factors. *Environ. Res.* **2020**, *190*, 109993. [CrossRef] [PubMed]
126. López-Bueno, J.A.; Navas-martín, M.A.; Linares, C.; Mirón, I.J.; Luna, M.Y.; Sánchez-Martínez, G.; Culqui, D.; Díaz, J. Analysis of the impact of heat waves on daily mortality in urban and rural areas in Madrid. *Environ. Res.* **2021**, *195*, 110892. [CrossRef] [PubMed]
127. Macintyre, H.; Heaviside, C.; Cai, X.; Phalkey, R. The winter urban heat island: Impacts on cold-related mortality in a highly urbanized European region for present and future climate. *Environ. Int.* **2021**, *154*, 106530. [CrossRef] [PubMed]
128. Gouveia, J.P.; Seixas, J. Unraveling electricity consumption profiles in households through clusters: Combining smart meters and door-to-door surveys. *Energy Build.* **2016**, *116*, 666–676. [CrossRef]
129. Pino-Mejías, R.; Pérez-Fargallo, A.; Rubio-Bellido, C.; Pulido-Arcas, J.A. Artificial neural networks and linear regression prediction models for social housing allocation: Fuel Poverty Potential Risk Index. *Energy* **2018**, *164*, 627–641. [CrossRef]
130. Bienvenido-Huertas, D.; Pérez-Fargallo, A.; Alvarado-Amador, R.; Rubio-Bellido, C. Influence of climate on the creation of multilayer perceptrons to analyse the risk of fuel poverty. *Energy Build.* **2019**, *198*, 38–60. [CrossRef]
131. Serrano-Jiménez, A.; Lizana, J.; Molina-Huelva, M.; Barrios-Padura, Á. Indoor environmental quality in social housing with elderly occupants in Spain: Measurement results and retrofit opportunities. *J. Build. Eng.* **2020**, *30*, 101264. [CrossRef]
132. Castaño-Rosa, R.; Barrella, R.; Sánchez-Guevara, C.; Barbosa, R.; Kyprianou, I.; Paschalidou, E.; Thomaidis, N.S.; Dokupilova, D.; Gouveia, J.P.; Kádár, J.; et al. Cooling degree models and future energy demand in the residential sector. A seven-country case study. *Sustainability* **2021**, *13*, 2987. [CrossRef]
133. Hornik, K.; Stinchcombe, M.; White, H. Multilayer feedforward networks are universal approximators. *Neural Netw.* **1989**, *2*, 359–366. [CrossRef]
134. Curry, B. Neural networks and seasonality: Some technical considerations. *Eur. J. Oper. Res.* **2007**, *179*, 267–274. [CrossRef]
135. Zhang, G.P.; Qi, M. Neural network forecasting for seasonal and trend time series. *Eur. J. Oper. Res.* **2005**, *160*, 501–514. [CrossRef]
136. Bandara, K.; Bergmeir, C.; Smyl, S. Forecasting across time series databases using recurrent neural networks on groups of similar series: A clustering approach. *Expert Syst. Appl.* **2020**, *140*, 112896. [CrossRef]
137. Alexander, D.L.J.; Tropsha, A.; Winkler, D.A. Beware of R2: Simple, unambiguous assessment of the prediction accuracy of QSAR and QSPR models. *J. Chem. Inf. Model.* **2015**, *55*, 1316–1322. [CrossRef] [PubMed]
138. Kvalseth, T.O. Cautionary note about r2. *Am. Stat.* **1985**, *39*, 279–285. [CrossRef]
139. Roth, M. Review of urban climate research in (sub)tropical regions. *Int. J. Climatol.* **2007**, *27*, 1859–1873. [CrossRef]
140. Schwarz, N.; Lautenbach, S.; Seppelt, R. Exploring indicators for quantifying surface urban heat islands of European cities with MODIS land surface temperatures. *Remote Sens. Environ.* **2011**, *115*, 3175–3186. [CrossRef]
141. Suomi, J. Extreme temperature differences in the city of Lahti, southern Finland: Intensity, seasonality and environmental drivers. *Weather Clim. Extrem.* **2018**, *19*, 20–28. [CrossRef]
142. Zhou, B.; Rybski, D.; Kropp, J.P. On the statistics of urban heat island intensity. *Geophys. Res. Lett.* **2013**, *40*, 5486–5491. [CrossRef]
143. Fu, P.; Weng, Q. Variability in annual temperature cycle in the urban areas of the United States as revealed by MODIS imagery. *ISPRS J. Photogramm. Remote Sens.* **2018**, *146*, 65–73. [CrossRef]
144. Lazzarini, M.; Molini, A.; Marpu, P.R.; Ouarda, T.B.M.J.; Ghedira, H. Urban climate modifications in hot desert cities: The role of land cover, local climate, and seasonality. *Geophys. Res. Lett.* **2015**, *42*, 9980–9989. [CrossRef]
145. Zhou, B.; Lauwaet, D.; Hooyberghs, H.; De Ridder, K.; Kropp, J.P.; Rybski, D. Assessing Seasonality in the Surface Urban Heat Island of London. *J. Appl. Meteorol. Climatol.* **2016**, *55*, 493–505. [CrossRef]

 *sustainability*

MDPI

Article

Indoor Air Quality in Naturally Ventilated Classrooms. Lessons Learned from a Case Study in a COVID-19 Scenario

Alberto Meiss *, Héctor Jimeno-Merino *, Irene Poza-Casado, Alfredo Llorente-Álvarez and Miguel Ángel Padilla-Marcos

RG Architecture & Energy, Universidad de Valladolid, 47014 Valladolid, Spain; irene.poza@uva.es (I.P.-C.); llorente@arq.uva.es (A.L.-Á.); miguelangel.padilla@uva.es (M.Á.P.-M.)
* Correspondence: alberto.meiss@uva.es (A.M.); hector.jimeno@uva.es (H.J.-M.)

Abstract: This paper describes the implementation of a series of ventilation strategies in a nursery and primary school from September 2020, when the government decided to resume the students' face-to-face activity in the middle of a COVID scenario. Air quality and hygrothermal comfort conditions were analysed before the pandemic and compared for different ventilation configurations in a post-COVID scenario. Ventilation strategies included the protocols issued by the Public Administration, while others were developed based on the typological configuration and use of the school. Results revealed that it is advisable to implement certain strategies that reduce the risk of infection among the occupants of the spaces, without a significant decrease in hygrothermal comfort. Given the importance of maintaining better IAQ in the future within classrooms, and regarding the pre-COVID situation, these strategies may be extended beyond this pandemic period, through a simple protocol and necessary didactic package to be assumed by both teachers and students of the centre.

Keywords: indoor air quality; COVID-19; educational buildings; natural ventilation

Citation: Meiss, A.; Jimeno-Merino, H.; Poza-Casado, I.; Llorente-Álvarez, A.; Padilla-Marcos, M.Á. Indoor Air Quality in Naturally Ventilated Classrooms. Lessons Learned from a Case Study in a COVID-19 Scenario. *Sustainability* **2021**, *13*, 8446. https://doi.org/10.3390/su13158446

Academic Editor: Tomonobu Senjyu

Received: 15 July 2021
Accepted: 27 July 2021
Published: 28 July 2021

Publisher's Note: MDPI stays neutral with regard to jurisdictional claims in published maps and institutional affiliations.

1. Introduction

The face-to-face return to classrooms in Spain in September 2020 during the global pandemic and after a long period of confinement opened up a debate in society about health security and air quality. The discussion was reinforced when the scientific community began to show evidence of greater transmission by aerosols than by fomites [1]. Poor indoor air quality (IAQ) was already an existing problem in naturally ventilated schools, but awareness was only raised after the COVID crisis. During the past school year, recommendations and protocols have been developed by different institutions and organisations to improve the ventilation performance of classrooms.

The objective of this study is the evaluation of the main protocols presented to improve natural ventilation systems (NVS) in schools, as well as the assessment of their possible adaptations to a post-COVID scenario, through the analysis of a case study.

Before the beginning of this global pandemic (February 2020), measurements were carried out in a nursery and primary school, which yielded worrying results: throughout a winter week of monitoring, the CO_2 concentration in classrooms exceeded 1000 ppm during 88.75% of the teaching time, reaching maximum values of 3628.8 ppm. These poor results are in line with those obtained in various previous studies (Table 1).

Inside classrooms, pupils and teachers are commonly the only sources of CO_2. Therefore, CO_2 is considered a good IAQ indicator, which shows the relationship between ventilation rate and occupancy.

Table 1. Comparison CO_2 values in different studies.

Reference	City Country	Year	Studied Classrooms	Ventilation System	Main Results
University of Burgos & PSBP [2]	Spain	2020	36	Natural Ventilation	More than 1000 ppm during 67.6% of the teaching time
Jørn Toftum et al. [3]	Denmark	2015	820	Natural & Mechanical	More than 1000 ppm during 66% of the teaching time
Almeida et al. [4]	Viseu (Portugal)	2017	76	Natural Ventilation	More than 2000 ppm during 25% of the teaching time. Maximum above 3000 ppm
Turanjanin [5]	Serbia	2014	5	Natural Ventilation	More than 1000 ppm during 50.0% of the teaching time. Maximum above 3600 ppm
Settimo et al. [6]	Rome (Italy)	2020	24	Not Specified	Daily values from 653 to 1352 ppm. Maximum average value of 2386 $\pm$ 480 ppm
Vassura et al. [7]	Bologna (Italy)	2015	2	Not Specified	Maximum average value of 3000 $\pm$ 1000 ppm

In addition, CO_2 has been associated with the presence of other pollutants [4,8]. The presence of pollutants like bioaerosols, Particle Matter (PM), or Total Volatile Organic Compounds (TVOCs) [9] is harmful to a pupil's health and productivity [10–12]. CO_2 must be taken into account as a pollutant, too, because, although not injurious at the levels that have been recorded in schools, it can cause adverse effects on the academic performance of the occupants [13].

TVOCs have diverse sources, mainly from the interior. These can come from cleaning products, construction materials or furniture, and, in the case of educational environments, from school materials such as glue, paint, etc. [14].

The approach to improve IAQ in schools is to increase ventilation rates. The first regulatory requirement in Spain for ventilation in schools dates from 1981 [15], which was endorsed in the "Regulation of thermal installations in buildings" (RITE) (1998) [16], and by its subsequent revision in 2007, currently in force with modifications [17].

Since in Castilla y León, 51% of public schools were built before 1980, and the vast majority of these centres lack ventilation systems, which involves addressing natural ventilation performance. This percentage is very similar to the Spanish average [18].

The requirements inside the classrooms are comparable between the different countries, according to Table 2.

Table 2. CO_2 concentration requirements inside the classrooms in different countries.

Region and Policy	CO_2 Concentration [ppm]
Europe [EN 16798-1:2020] [1]	950 [2]
Finland [Decree Indoor Climate and Ventilation 1009 (2017)]	1200 [2]
Germany [DIN 1946-4 (2005)]	1500
Portugal [RECS (2013)]	1250
Spain [RITE (2007)]	900 [2]
United Kingdom [Building Bulletin 101 (2018)]	1500 [3]
USA [ASHRAE 62.1 (2019)]	1100 [2]

[1] Value for Category I (educational buildings—new buildings and renovations). [2] Outdoor air concentration was assumed as 400 ppm. [3] Value obtained for up to 20 min.

The uncertainty regarding the real performance of ventilation systems in educational centres has promoted different studies on the improvement of natural ventilation.

Almeida et al. (2015) compared the IAQ between two refurbished and two non-refurbished classrooms for two months in Spring. All of them had a central heating system with hot-water radiators as terminal units. A mechanical ventilation system (MVS) provided with CO_2 and temperature sensors (indoors and outdoors as a reference value) was implemented in the refurbished classrooms. The results obtained revealed that the natural ventilation system was not able to provide adequate IAQ for the whole school day. The CO_2 concentration levels were over 1500 ppm for 20% of the time. On the other hand, while using an MVS a good IAQ was maintained [19].

Also in a mild-climate location, Fernández-Agüera et al. (2019) [20] found that a large part of the schools lacked ventilation systems. Therefore, the air renewal of the classrooms was achieved by means of uncontrolled airflow through the building envelope (air infiltration) or the manual opening of windows. The data showed that, when the windows remained closed, the CO_2 concentration reached values over 1000 ppm during 89.3% of the time.

Vasella et al. (2021) [21] studied the impact of an NVS protocol implemented in a hundred schools in Switzerland for four days during the cold season with outdoor temperatures below 15 °C (please note that these are similar climate conditions to the ones of the present study). The protocol consisted of the brief opening of doors and windows between different classes and for a longer period at the lunch break when the students had to leave the classroom for its ventilation. The measures implemented included an application to calculate the duration of the brief apertures, and teaching materials to promote the importance of IAQ and ventilation. The protocol implemented reached CO_2 concentration levels under 1400 ppm during 70% of the school time, whereas those levels were only achieved during 30% of the time before its implementation, and the mean temperature was between 19.9 °C and 20.5 °C.

Similar research in the Netherlands carried out in 2008 [22] demonstrated that when protocols have been implemented to raise awareness of the problem among students and teachers, it is possible to achieve a necessary longer-lasting effect. Therefore, before the intervention, the CO_2 concentration exceeded 1000 ppm for 64% of the school day, whereas after the intervention, ventilation was significantly improved even though the CO_2 concentration still exceeded 1000 ppm for more than 40% of the school day.

Thus, there is enough existing evidence to indicate that it is necessary to control IAQ in schools. However, it is also necessary to evaluate what success it could achieve, and how it could affect the thermal comfort in schools in Castilla y León.

In a scenario in which health had to be prioritised, certain ventilation protocols that compromised thermal comfort were reappraised. To determine the scenarios subject to study, the main mandatory protocols and guidelines published to date were taken into account:

- On 19 June 2020, the government of Castilla y León issued "Prevention and organisation protocol for the return to school activity in the educational centres of Castilla y León for the school year 2020/2021" [23];
- In June 2020, the Harvard T.H. Chan School of Public Health published the guide "Healthy Schools: Risk reduction strategies for reopening schools" [24], and in August 2020 "5-step guide to checking ventilation rates in classrooms" [25];
- In October 2020, CSIC-IDAEA released the "Guide to classroom ventilation," which has some common ground with the one just mentioned. In December 2020 this guide was upgraded with results obtained from real cases [14];
- In February 2021, the report of CSIC-LIFTEC "Continuous Ventilation vs. Flashing Ventilation" was published [26].

2. Methodology

2.1. Site and Building

The case of study is a primary public school in Valladolid (Castilla y León, Spain). The building is placed in a plot mostly occupied by the playground, and it is next to other public buildings to the north and south sides, a park to the east and a four-lane avenue with medium-high traffic volume to the west (Figure 1). Valladolid has a Continental Mediterranean climate with cold winters and minimum temperatures below zero

(Csb-Temperate, dry and temperate summer). As a consequence, maintaining the indoor temperature efficiently becomes an important factor.

Figure 1. Aerial view of the school from Google Earth Pro (2021).

The building, which was built in 1980, has a construction system that was widespread in the construction of public educational buildings built in the same decade: a reinforced concrete structure and vertical envelope composed of brick masonry, a non-ventilated air chamber (insulated only on some occasions) and single-hollow brick as the inner layer. It has a ground floor and two additional floors, with a central corridor and classrooms on each side (oriented to north and south). Each classroom (Figure 2) has a rectangular floor area of 60 m^2 and is 2.85 m in height. The classrooms have two doors to the corridor, with a panel of adjustable methacrylate slats over them, and four tilt-and-turn PVC windows with integrated shutters to the exterior (the original exterior windows were recently replaced).

Figure 2. 360° view from a tested classroom.

The building is naturally ventilated, and the heating system has a central boiler and aluminium radiators, with uninsulated pipes through the classrooms.

The usual class schedule is Monday through Friday from 9:00 a.m. to 2:00 p.m., with an intermediate break time of 30 min.

2.2. Test Design

Different parameters that define IAQ in classrooms were compared in different ventilation scenarios. In the first phase (pre-COVID-19), IAQ parameters before the pandemic under no ventilation protocol were collected. Next, scenarios A and B, which emerged from the protocol prescribed by the public administration in Castilla y León [21], were assessed. Finally, alternative scenarios (C, D and D') were defined according to the guidelines proposed by the Harvard T.H. Chan School of Public Health and CSIC [22–24]. At the same time, hygrothermal comfort conditions for the different ventilation configurations were analysed, considering their viability in a post-COVID time.

For each test, measurements were carried out simultaneously in two facing representative classrooms (Figure 3), with the aim of evaluating the impact of cross ventilation. In this regard, cross-ventilation was determined from the data monitored in the corridor from the simultaneous opening of doors and windows in facing classrooms. This resulted in a significant increase in the ventilation flow by the pressure gradient between the windward and leeward façades.

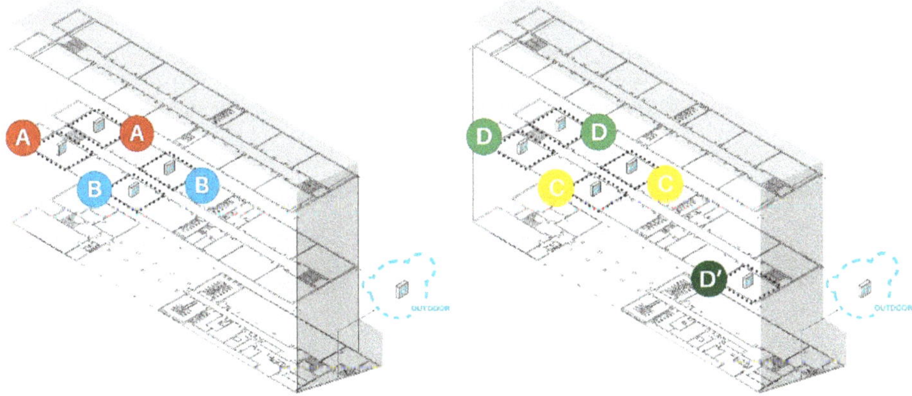

Figure 3. Scenario locations and classrooms within the building.

All tests were carried out during the cold season. The data collection for the pre-COVID-19 scenario was carried out for 5 school days (Monday–Friday), 3 days (Wednesday–Friday) for scenarios A and B and 5 school days (Monday–Friday) for scenarios C, D and D' (Table 3). During all the test phases the occupation of the classrooms was constant.

Table 3. Organisation of classrooms and test protocols.

Classroom	Pre-COVID-19	Scenario A	Scenario B	Scenario C	Scenario D	Scenario D'
Preschool						X
Level 1°A	X	X			X	
Level 1°C		X			X	
Level 1°D			X	X		
Level 2°A	X					
Level 2°C	X		X	X		

Other sensors took measures outdoors and in the corridor, respectively.

- Scenario A: IAQ was monitored in the classrooms under the application of the government protocol. This regulation established protection measures (mandatory use of a face mask for all students over 6 years old), safety distancing (chairs at a minimum distance of 1.5 m, preventing the occupants from sitting facing each other) and the ventilation of classrooms. Mandatory ventilation had to take place between 10 and 15 min before the arrival of the students, at the end of each school class (5 min after a 55-min non-ventilated period), during the break (30 min) and at the end of the day. (The teachers and students were unaware of the nature of the sampling carried out on these days.);
- Scenario B: a similar situation to the previous one, but CO_2 sensors were provided to teachers so that, in light of the protocol, they could act according to their criteria in case excessive concentrations were detected during the tests;
- Scenario C: the continuous opening of the windows was tested. Thus, the four windows in each classroom were kept open in an oscillating position (approximately 0.18 m^2 of free surface each) throughout the school day. Moreover, to maintain cross-ventilation, at least one of the two doors to the hall was open. During the break, two windows were completely open (0.72 m^2 of free surface each), while the other two remained in the aforementioned position. In addition, it was possible to assess not only the suitability of the indoor air renewal but also the operability of the situation. Likewise, the teachers had CO_2 sensors in case excessive concentrations were detected during the tests;
- Scenario D: the windows of the classroom were opened at specific times. All four windows (0.72 m^2 of free surface each) and both doors (1.65 m^2 of free surface each) were completely opened for 5 min after a 25-min non-ventilated period. For example, if the class started at 9:00 a.m., users opened the windows at 9:25 and closed them at 9:30 a.m., and so on for the whole school day. During the break, all the windows were kept open in an oscillating position (0.18 m^2 of free surface each) and doors were completely opened. The operation was carried out simultaneously in the facing classrooms to force cross-ventilation;
- Scenario D': a slight variant of Scenario D. In this case, the scheme of apertures was the same as Protocol D, but the classrooms were ventilated constantly through the small slats over the doors (0.35 m^2 of free surface each), while the doors (1.65 m^2 of free surface each) were kept closed. This means that there was not important cross-ventilation.

In all cases, the opening scheme could be modified at the discretion of the teachers and such circumstances were registered.

2.3. IAQ Monitoring

The sensors used were AirQualityEgg, which measure several IAQ conditions, specifically: air temperature, relative humidity (RH), CO_2, TVOC, PM_{10}, $PM_{2.5}$ and $PM_{1.0}$. The systematic measurement error (bias) of the different sensors integrated were:

- CO_2: ± 30 ppm/$\pm 3\%$ of the measured value;
- TVOC: Sensor IAQ—Core Indoor Air Quality (Resolution 16 bits);
- PM: $\pm 10\%$;
- Humidity sensor: $\pm 2\%$RH;
- Temperature sensor: $\pm 0.3\ ^\circ$C.

Measurements were taken every five minutes from the previous hour, before the academic activity, up to one hour after its completion. The position of the measurement devices within the classroom was determined by previous tests, to limit distortions produced by the occupancy, academic material, blackboards, windows and doors. Sensors were kept between 1 and 2 m away from the area where the students were, and at least 1 m away from other possible disturbance sources. Regarding height, negligible variations of less than 3% in CO_2 and pollutant levels were observed between the breathing plane of the students in a sitting position (1.20 m) and 1 m above (2.20 m above the floor level).

2.4. IAQ Limit Values

The limits established for the different pollutants were determined based on the criteria established by different regulations and guidelines:

- Operative temperature: between 21 °C and 23 °C during the heating season [17];
- RH: between 40% and 50% during the heating season [17];
- CO_2: 900 ppm (absolute value, i.e., 500 ppm over outdoor CO_2 concentration), defined as InDoor Air level (IDA) 2 for educational buildings [17];
- PM: since 2005, the World Health Organization (WHO) recommends a daily mean concentration of $PM_{2.5}$ below 25 µg/m^3, and below 50 µg/m^3 in the case of $PM_{10.0}$ [27];
- TVOC: 1.00 mg/m^3 or 500 ppb are considered adequate limits [28].

3. Results

The results obtained in this study are structured in three phases. Firstly, the data obtained during the winter season in 2020, in a pre-COVID-19 situation, are shown. Afterwards, Phases 1 and 2, with the data of the different ventilation protocols evaluated, during the winter season in 2021 and in a COVID-19 scenario, are presented (Table 4).

In the scenario pre-COVID-19, the monitoring results obtained in the classrooms (Level 2°C, 1°A, 2°A and Pre-school) revealed very poor values, achieving a maximum CO_2 concentration gradient indoors of 4025 ppm, more than six times above the limit set for RITE [17]. Taking into account the CO_2 concentration, its level was out of the normative range between 81% and 93% of the school time. The TVOC concentration was out of the range between 3% and 50% of the time, with a maximum value of 767 ppm reached in Level 2°A. In all cases, the level of PM was adequate under the maximum recommended levels. The ventilation was quite low, and cross-ventilation was close to zero. The rare opening of windows entailed that the only air change occurred through air infiltration. The lack of ventilation also caused the increase in temperature during the school day and poor IAQ.

Furthermore, cross-ventilation was an active strategy within scenarios C and D. In these scenarios, the classrooms maintained an average comfort temperature throughout the day, but there were significant periods with out-of-range conditions. In all cases, the mean CO_2 concentration level during the school day was within the adequate range. However, there were short periods in which CO_2 levels were above the maximum recommended values, namely for scenarios A (7.10%) and D (6.23%). This implies that the classroom had inadequate ventilation rates for 20 min per school day. PM was not a problem in any case. Instead, TVOC levels were over the recommended level of 500 ppm in certain moments, in a small percentage.

Finally, it is worth mentioning that during the performance of the tests, the conditions of the outdoor environment were continuously monitored (Table 5). The average outdoor temperature and the temperature differential between the hour before the start of the activity and the average temperature during the day were practically the same. Of particular importance are the CO_2 values outdoors since they serve as a reference in most of the regulations based on the concentration gradient between the inside and the outside.

Table 4. Results.

Test	Occupation	Ventilation			Temperature			RH		CO_2 INDOOR			PM			TVOC	
	$n°$	Mean Window Opening [m^2]	Mean Door Opening [m^2]	Cross-Ventilation	Max. [°C]	Min. [°C]	Time out of Range	Mean	Time out of Range	Max. [ppm]	Mean [ppm]	Time out of Range	Mean $PM_{2.5}$ [$\mu g/m^3$]	Mean PM_{10} [$\mu g/m^3$]	Time out of Range	Mean [ppb]	Time out of Range
pre-COVID-19 Level 2°C	22	0.01	0.01	0.1%	25.4	20.7	1.4%	45.5%	15%	3764	2264	93%	5.5	7.2	0.0%	485	50%
pre-COVID-19 Level 1°A	21	0.13	0.48	0.1%	26.5	21.4	0.0%	38.3%	58%	2486	1422	81%	9.6	10.7	0.0%	302	3%
pre-COVID-19 Level 2°A	23	0.03	0.14	0.1%	25.5	21.1	0.0%	46.0%	9%	4025	2232	93%	4.8	5.8	0.0%	483	38%
pre-COVID-19 Level Preschool	21	0.06	0.17	0.1%	26.6	21.5	0.0%	40.6%	47%	3629	1918	88%	7.2	8.7	0.0%	416	25%
Scenario A Level 1°A & C	18	0.98	0.92	46%	22.4	17.5	65.8%	37.6%	87%	1255	628	7.1%	11.5	14.1	0.0%	294	10%
Scenario B Level 1°D & 2°A	19	0.97	0.79	18%	25.2	17.5	41.8%	35.4%	93%	1044	577	1.9%	9.6	11.9	0.0%	298	10%
Scenario C Level 1°D & 2°A	20	1.02	1.18	100%	24.5	19.2	18.7%	38.3%	81%	1011	638	0.3%	14.6	16.9	0.0%	287	3%
Scenario D Level 1°A & C	18	0.52	0.43	79%	23.0	18.8	46.6%	39.7%	53%	1047	707	3.4%	15.3	17.4	2.5%	320	9%
Scenario D' Level Preschool	17	0.85	0.55	0.0%	24.5	18.9	11.0%	37.7%	89%	1200	668	6.2%	13.5	15,5	0.0%	319	6%

Table 5. Outdoor conditions for each test Phase.

Week	Day	Temperature [°C]	RH [%]	Wind Speed [km/h]	CO_2 [ppm]	$PM_{2.5}$ [$\mu g/m^3$]	PM_{10} [$\mu g/m^3$]	TVOC [ppb]
Phase pre-COVID February 2020	Mon.	12.1	68.6	6.8	440	5.3	5.7	227
	Tues.	7.1	60.7	3.9	443	7.2	7.8	281
	Wed.	5.2	60.7	3.9	450	16.0	17.3	405
	Thurs.	5.3	57.6	2.9	452	24.7	27.2	424
	Fri.	9.0	70.9	2.8	442	25.8	28.1	202
	Mean	7.8	63.9	4.1	446	15.8	17.2	308
Scenarios A & B February 2021	Wed.	11.3	59.9	12.0	426	10.7	14.2	202
	Thurs.	11.9	55.6	6.1	420	9.4	13.9	193
	Fri.	8.8	62.5	8.0	457	20.1	23.26	344
	Mean	10.7	59.4	8.7	433	13.4	17.1	246
Scenarios C, D & D′ February 2021	Mon.	10.6	59.6	6.9	439	20.2	24.8	227
	Tues.	10.8	63.2	5.0	434	23.0	27.3	225
	Wed.	9.9	63.4	4.0	428	19.5	22.4	221
	Thurs.	12.2	59.4	3.0	440	21.7	25.7	230
	Fri.	9.6	65.4	6.0	421	27.0	32.5	182
	Mean	10.6	62.2	4.9	432	22.3	26.6	217

Inquiry to the Teachers

After the tests, a brief survey was sent to the ten teachers of the classrooms studied, with two questions related to their experience during the process:

- Question 1: In a COVID-19 scenario, would it be possible and realistic to keep this ventilation protocol in the classrooms?
- Question 2: In a post-COVID-19 scenario, would it be possible and realistic to introduce a ventilation protocol which is less restrictive into your daily routine for the long term?

They could also make additional comments or clarifications (Table 6).

Table 6. Survey to the teachers after the tests (1 teacher per class, 2 teachers per scenario).

Scenario	Question 1		Question 2		Comments
	Yes	No	Yes	No	
A	2 100%	0 0%	2 100%	0 0%	
B	2 100%	0 0%	2 100%	0 0%	
C	2 100%	0 0%	2 100%	0 0%	
D	0 0%	2 100%	2 100%	0 0%	Rigid protocol to diary routine
D′	2 100%	0 0%	2 100%	0 0%	Cold/Perception of poor ventilation

4. Discussion

In pre-COVID cases, the lack of ventilation is evident. In some situations, there is almost none, except when the windows and doors are rarely opened during the midday break. Only in the scenario pre-COVID-19 Level 1°A was the intention of a routine indoor air renewal perceived; thus, even if ventilation is scarce, it is enough to maintain adequate TVOC concentration for most of the time. In this classroom, the average indoor CO_2 was 34% below the other classrooms, but even so, during 81% of the week's school time, or for 4 out of the 5 teaching hours, CO_2 levels exceeded the limit set by Spanish regulations.

From the tests performed on the scenarios (A–D′), the most important aspects of the impact of ventilation on the monitored contaminants were analysed:

- CO_2: in all scenarios, the average CO_2 concentration was low, well below the range established by regulations (around 900 ppm), so good IAQ conditions were guaranteed throughout the school day. Scenario C guaranteed better IAQ since the time out of the quality standards was negligible;
- Scenarios A and D′ were the most unfavourable in terms of time out of range and maximum values obtained. Although the time with CO_2 concentration levels out of range did not exceed 19 min, which could be acceptable, the maximum values recorded rose up to 1255 ppm, 33.3% above the limit, which makes it not a recommended solution.
- TVOC: comparing the pre-COVID-19 scenario and COVID-19 scenarios, mean concentration levels during school hours in the classrooms improved from 421 ppm to 320 ppm, and from 29.0% of the out-of-range time to 10%. Therefore, overall, it can be said that the proposed ventilation scenarios succeeded to register adequate TVOC concentration in the classrooms.
- In Scenario C, with continuous ventilation, the average TVOC concentration was lower than the average registered in other scenarios and, also, the time out of range was notably lower. This may be due to the fact that, in cases where ventilation is rarely promoted, there were long periods when TVOCs accumulated, especially when school materials that are VOC sources (markers, paints and other handicraft materials) were used.
- PM: this did not entail a problem within the classrooms. The presence of PM inside the classrooms was conditioned by the outdoor concentrations. Despite the fact that aerosols are vectors of SARS-CoV2, the presence of these does not necessarily imply a greater probability of infection since the external particles are assumed not contaminated and the transmission is carried out through the human-produced aqueous bioaerosols.

Regarding the indoor temperature conditions throughout the school day, it was observed that scenarios C and D′ were the most favourable, with values of just 18.7% and 11.0% of the time outside the acceptable temperature range.

One last aspect to consider for any natural ventilation strategy to be valid is its feasibility to be lasting over time. In other words, it must be compatible with the daily routine of the classroom. The surveys carried out revealed that the punctual opening entailed a great difficulty because, in order to be effective (cross-ventilation), coordination between two classrooms was needed. Therefore, it would be a solution with little scope. Additionally, to achieve the durability of the protocol, Geelen et al. (2008) [22] pointed out the need for a didactic package attached to the scenario of application, for both a COVID-19 or post-COVID-19 scenario. This may have failed in Scenario A, as reflected in the results.

This study was limited to a case study, so it would be necessary to evaluate its applicability to other educational centres in order to achieve a larger sample and representativeness.

5. Conclusions

Despite the limitations that a case study supposes, the proposed methodology and proceedings can serve as guidelines for a post-COVID-19 scenario.

Considering IAQ, thermal comfort and the practicality of the scenarios, scenario C is the most effective for this case study. Thus, the mixing model, which ensures the removal of pollutants, is guaranteed by continuous cross-ventilation. It can be reached using different opening levels, depending on the use and occupation of the classroom, which contributes to acquiring an adequate air quality for the use of the space. Air flow regulation reduces the thermal effects of the new air coming from outdoors.

A similar air flow can be obtained by fully opening doors and windows in designated moments of the day. However, there is both a thermal loss and an absence of cross-ventilation while teaching. These conditions promote an air quality fluctuation value, while a constant value is preferred for these kinds of spaces. This can be observed in the application of Scenario B, with similar ventilation values but 82% lower cross-ventilation, where adequate thermal comfort results were not achieved.

As demonstrated in Scenario B and Scenario D, having specific openings set was only effective if there was cross-ventilation. As a result, if the layout of the classrooms require coordination between them, it cannot be considered functional since its application over time would be compromised.

Overall, two types of protocols are suggested to be applied in naturally ventilated classrooms: one for health emergency scenarios, in which higher ventilation rates are required, and another for non-emergency situations, in which ventilation rates are not that demanding.

Finally, all these natural ventilation scenarios are influenced by the pollutants present outdoors. This aspect should contribute to defining criteria to operate the openings by the constant monitoring of those pollutants. In this case study, high-pollutant values were not detected, neither in PMs nor TVOC. For instance, it was possible to verify that the indoor concentration of pollutants is ruled by a proportional relationship with the values registered outdoors.

Author Contributions: Conceptualisation, A.M., H.J.-M. and I.P.-C.; methodology, A.M., H.J.-M. and I.P.-C.; formal analysis, A.M., A.L.-Á. and M.Á.P.-M.; investigation, H.J.-M.; resources, A.L.-Á.; data curation, H.J.-M.; writing—original draft preparation, H.J.-M.; writing—review and editing, A.M. and I.P.-C.; visualisation, A.M.; supervision, I.P.-C.; project administration, A.L.-Á.; funding acquisition, M.Á.P.-M. All authors have read and agreed to the published version of the manuscript.

Funding: This research "Metodologías de estudio y estrategias de mejora de eficiencia energética, confort y salubridad de centros educativos en Castilla y León. Investigación básica" was funded by JUNTA DE CASTILLA Y LEÓN. CONSEJERÍA DE EDUCACIÓN, grant number VA026G19 (2019–2021).

Institutional Review Board Statement: Not applicable.

Informed Consent Statement: Not applicable.

Conflicts of Interest: The authors declare no conflict of interest.

References

1. Morawska, L.; Milton, D.K. It Is Time to Address Airborne Transmission of Coronavirus Disease 2019 (COVID-19). *Clin. Infect. Dis.* **2020**, *71*, 2311–2313. [CrossRef] [PubMed]
2. Plataforma de Edificación Passivhaus; Universidad de Burgos Proyecto de Monitorización de colegios. Estudio Completo e Informe de Conclusiones. 2020; pp. 1–261. Available online: http://plataforma-pep.s3.amazonaws.com/documents/documents/000/000/873/original/PEP._Estudio_de_monitorizaci%C3%B3n_de_colegios_completo_F.PDF?1602758126 (accessed on 10 July 2021).
3. Toftum, J.; Kjeldsen, B.U.; Wargocki, P.; Menå, H.R.; Hansen, E.M.N.; Clausen, G. Association between classroom ventilation mode and learning outcome in Danish schools. *Build. Environ.* **2015**, *92*, 494–503. [CrossRef]
4. Almeida, R.M.S.F.; Pinto, M.; Pinho, P.G.; de Lemos, L.T. Natural ventilation and indoor air quality in educational buildings: Experimental assessment and improvement strategies. *Energy Effic.* **2017**, *10*, 839–854. [CrossRef]
5. Turanjanin, V.; Vučićević, B.; Jovanović, M.; Mirkov, N.; Lazović, I. Indoor CO_2 measurements in Serbian schools and ventilation rate calculation. *Energy* **2014**, *77*, 290–296. [CrossRef]
6. Settimo, G.; Indinnimeo, L.; Inglessis, M.; De Felice, M.; Morlino, R.; Di Coste, A.; Fratianni, A.; Avino, P. Indoor air quality levels in schools: Role of student activities and no activities. *Int. J. Environ. Res. Public Health* **2020**, *17*, 6695. [CrossRef] [PubMed]
7. Vassura, I.; Venturini, E.; Bernardi, E.; Passarini, F.; Settimo, G. Assessment of indoor pollution in a school environment through both passive and continuous samplings. *Environ. Eng. Manag. J.* **2015**, *14*, 1761–1770. [CrossRef]

8. Chatzidiakou, L.; Mumovic, D.; Summerfield, A. Is CO_2 a good proxy for indoor air quality in classrooms? Part 1: The interrelationships between thermal conditions, CO2 levels, ventilation rates and selected indoor pollutants. *Build. Serv. Eng. Res. Technol.* **2015**, *36*, 129–161. [CrossRef]

9. Becerra, J.A.; Lizana, J.; Gil, M.; Barrios-Padura, A.; Blondeau, P.; Chacartegui, R. Identification of potential indoor air pollutants in schools. *J. Clean. Prod.* **2020**, *242*. [CrossRef]

10. Gilliland, F.D.; Berhane, K.; Rappaport, E.B.; Thomas, D.C.; Avol, E.; Gauderman, W.J.; London, S.J.; Margolis, H.G.; McConnell, R.; Islam, K.T.; et al. The effects of ambient air pollution on school absenteeism due to respiratory illnesses. *Epidemiology* **2001**, *12*, 43–54. [CrossRef]

11. Mendell, M.J.; Heath, G.A. Do indoor pollutants and thermal conditions in schools influence student performance? A critical review of the literature. *Indoor Air* **2005**, *15*, 27–52. [CrossRef]

12. Mujan, I.; Anđelković, A.S.; Munćan, V.; Kljajić, M.; Ružić, D. Influence of indoor environmental quality on human health and productivity—A review. *J. Clean. Prod.* **2019**, *217*, 646–657. [CrossRef]

13. Satish, U.; Mendell, M.J.; Shekhar, K.; Hotchi, T.; Sullivan, D.; Streufert, S.; Fisk, W.J. Is CO_2 an indoor pollutant? direct effects of low-to-moderate CO2 concentrations on human decision-making performance. *Environ. Health Perspect.* **2012**, *120*, 1671–1677. [CrossRef] [PubMed]

14. Minguillón, M.C.; Querol, X.; Felisi, J.M.; Garrido, T. *Guía para Ventilación en Aulas*; Consejo Superior de Investigaciones Científicas: Madrid, Spain, 2020.

15. Presidencia del Gobierno. Orden de 16 de julio de 1981 por la que se aprueban las instrucciones técnicas complementarias denominadas IT.IC, con arreglo a lo dispuesto en el Reglamento de Instalaciones de Calefacción, Climatización y Agua Caliente Sanitaria. *Boletín Oficial del Estado*, 13 August 1981; 18578–18620.

16. Ministerio de la Presidencia. RITE 1998. Reglamento de Instalaciones Térmicas en los Edificios. *Boletín Oficial del Estado*, 5 August 1998; 26585–26587.

17. Ministerio de la Presidencia. RITE 2007. Reglamento de instalaciones térmicas en los edificios. *Boletín Oficial del Estado*, 29 August 2007; 35931–35984.

18. Gobierno de España. Actualización 2020 de la Estrategia a largo plazo para la Rehabilitación Energética en el Sector de la Edificación en España. *ERESEE*, 18 June 2020; 376.

19. Almeida, R.M.S.F.; De Freitas, V.P. IEQ assessment of classrooms with an optimized demand controlled ventilation system. *Energy Procedia.* **2015**, *78*, 3132–3137. [CrossRef]

20. Fernández-Agüera, J.; Campano, M.Á.; Domínguez-Amarillo, S.; Acosta, I.; Sendra, J.J. CO_2 concentration and occupants' symptoms in naturally ventilated schools in mediterranean climate. *Buildings* **2019**, *9*, 197. [CrossRef]

21. Vassella, C.C.; Koch, J.; Henzi, A.; Jordan, A.; Waeber, R.; Iannaccone, R.; Charrière, R. From spontaneous to strategic natural window ventilation: Improving indoor air quality in Swiss schools. *Int. J. Hyg. Environ. Health* **2021**, *234*. [CrossRef] [PubMed]

22. Geelen, L.M.J.; Huijbregts, M.A.J.; Ragas, A.M.J.; Bretveld, R.W.; Jans, H.W.A.; Van Doorn, W.J.; Evertz, S.J.C.J.; Van Der Zijden, A. Comparing the effectiveness of interventions to improve ventilation behavior in primary schools. *Indoor Air* **2008**, *18*, 416–424. [CrossRef] [PubMed]

23. Consejería de Educación. Junta de Castilla y León. Protocolo de Prevención y Organización del Regreso a la Actividad Lectiva en los Centros Educativos de Castilla y León para el Curso Escolar 2020/2021. 2020; pp. 1–19. Available online: https://www.educa.jcyl.es/es/institucional/crisis-coronavirus/crisis-coronavirus-protocolos-resoluciones-guias/cursos-anteriores/protocolo-prevencion-organizacion-regreso-actividad-lectiva (accessed on 10 July 2021).

24. Jones, E.; Young, A.; Clevenger, K.; Salimifard, P.; Wu, E.; Luna, M.L.; Lahvis, M.; Lang, J.; Bliss, M.; Azimi, P.; et al. Healthy schools: Risk reduction strategies for reopening schools. *Harvard TH Chan Sch. Public Health Healthy Build. Program.* **2020**. [CrossRef]

25. Joseph, A.; Jack, S.; Emily, J.; Jose, C.-L. *FOR HEALTH 5 Step Guide to Measuring Ventilation Rates in Classrooms*; Hardvard T.H. Chan: Boston, MA, USA, 2020.

26. Muelas, Á.; Pina, A.; Remacha, P.; Tizné, E.; Ballester, J.; CSIC, Universidad de Zaragoza. Ventilación Natural en las Aulas. Ventilación Continua vs. Intermitente. 2021. Available online: https://stecyl.net/wp-content/uploads/2021/02/VentilacionContinua_vs_Intermitente.pdf (accessed on 10 July 2021).

27. World Health Organization (WHO). *Guías de Calidad del aire de la OMS Relativas al Material Particulado, el Ozono, el Dióxido de Nitrógeno y el Dióxido de Azufre*; World Health Organization (WHO): Geneva, Switzerland, 2005.

28. Ahmed Abdul-Wahab, S.A.; En, S.C.F.; Elkamel, A.; Ahmadi, L.; Yetilmezsoy, K. A review of standards and guidelines set by international bodies for the parameters of indoor air quality. *Atmos. Pollut. Res.* **2015**, *6*, 751–767. [CrossRef]

 *sustainability*

Article

MgO-Based Cementitious Composites for Sustainable and Energy Efficient Building Design

Serenay Kara [1], Savas Erdem [1,*] and Roberto Alonso González Lezcano [2]

[1] Department of Civil Engineering, Faculty of Engineering, Istanbul-Cerrahpasa University, 34200 İstanbul, Turkey; serenaykara1@gmail.com
[2] Architecture and Design Department, Escuela Politécnica Superior, Universidad CEU San Pablo, 28040 Madrid, Spain; rgonzalezcano@ceu.es
* Correspondence: savas.erdem@istanbul.edu.tr

Abstract: Concrete made with Portland cement is by far the most heavily used construction material in the world today. Its success stems from the fact that it is relatively inexpensive yet highly versatile and functional and is made from widely available raw materials. However, in many environments, concrete structures gradually deteriorate over time. Premature deterioration of concrete is a major problem worldwide. Moreover, cement production is energy-intensive and releases a lot of CO_2; this is compounded by its ever-increasing demand, particularly in developing countries. As such, there is an urgent need to develop more durable concretes to reduce their environmental impact and improve sustainability. To avoid such environmental problems, researchers are always searching for lightweight structural materials that show high performance during both processing and application. Among the various candidates, Magnesia (MgO) seems to be the most promising material to attain this target. This paper presents a comprehensive review of the characteristics and developments of MgO-based composites and their applications in cementitious materials and energy-efficient buildings. This paper starts with the characterization of MgO in terms of environmental production processes, calcination temperatures, reactivity, and micro-physical properties. Relationships between different MgO composites and energy-efficient building designs were established. Then, the influence of MgO incorporation on the properties of cementitious materials and indoor environmental quality was summarized. Finally, the future research directions on this were discussed.

Keywords: MgO-based cement; sustainability; energy efficiency; sustainable materials; green architecture

Citation: Kara, S.; Erdem, S.; Lezcano, R.A.G. MgO-Based Cementitious Composites for Sustainable and Energy Efficient Building Design. *Sustainability* **2021**, *13*, 9188. https://doi.org/10.3390/su13169188

Academic Editor: Marc A. Rosen

Received: 6 June 2021
Accepted: 13 August 2021
Published: 16 August 2021

Publisher's Note: MDPI stays neutral with regard to jurisdictional claims in published maps and institutional affiliations.

1. Introduction

The adage "all old is fresh again" surely applies to the existing condition of magnesia-based cements (MgO). The worldwide building materials sector, which has been traditionally focused on a wide variety of materials suitable for local requirements and individual uses, became almost a monoculture based on the use of Portland cement (PC) during the latter half of the twentieth century, with the other materials largely overlooked. Because of the fast growth of the building sector, the production of Portland cement and natural aggregates has risen at an unprecedented rate. In fact, the building industry required around 40 billion tons of aggregates and 4 billion tons of cement in 2014 [1,2]. As a result, mostly during the manufacturing of these products, a large volume of carbon dioxide (CO_2) is emitted into the atmosphere. For instance, to generate 1 ton of cement, 125 kW of energy is needed, as well as 0.89 tons of CO_2 emissions into the atmosphere [3,4]. Implementing renewable materials into concrete, such as fly ash (FA), silica fume, slag, metakaolin, and industrial byproducts, is one way to address this problem [5–10]. Incorporating MgO into concrete is another choice for making it more natural. Cements with a high MgO content have become increasingly common in recent years, owing to the concern about climate change and the need to reduce CO_2 emissions associated with the manufacture of traditional Portland cements. Some scholars assume that cements with a high MgO material can be made with lower CO_2 emissions [11]. Other scholars assume that by trapping ambient CO_2 and converting it to magnesium crystals, it is possible to make cement with a positive CO_2 balance (carbonates and hydroxycarbonates). Carbonation of MgO is defined as the formation of magnesite from MgO as a result of the absorption of CO_2 [11].

In such pyrolysis and reaction situations, the use of MgO will reduce thermal shrinkage [12,13]. It is also possible to decrease the expense of concrete by eliminating the need for expensive curing steps and speed up the manufacturing process by continually casting concrete without the need for too many cold joints [14]. However, environmental concerns were the driving force behind the production and increasing capacity of MgO-based cements. Because of the need for MgO production by cooler altitudes (relative to the desired rate for processing $CaCO_3$ in ordinary Portland cement (OPC)), as well as the fuel savings related to low temperatures, many people see MgO-based cements as a key component of the future of environmentally sustainable cement manufacturing. Natural resources such as soil, stone, and timber are appropriate building materials for low carbon emissions and footprints and recycling or reuse potential. Natural materials that are unprocessed or minimally processed have limitations, notably in terms of strength and durability. Energy is expended in the processing and transportation of raw materials, resulting in carbon emissions. To reduce carbon emissions, developing technologies that allow building materials and products to be made with the least amount of energy [15]. The building industry is interested in the development of carbon-neutral cementation pathways as a significant challenge. Capturing CO_2 released during the calcination of limestone, and perhaps reusing it, is a very appealing technique to attain this goal. To that purpose, this research investigated the importance of reaction parameters, including time, temperature, and pressure, affect the rate of $Ca(OH)_2$ carbonation under both liquid and supercritical CO_2 exposure [16]. Similarly, MgO's capacity to absorb carbon dioxide from the air to shape a range of carbonate and hydroxycarbonate blends are on point with "carbon neutral" cements, which can absorb almost as much CO_2 over their lifespan as they emit during manufacturing. These two intertwined factors have sparked a recent surge in research and business involvement in MgO-based cements.

This paper starts with the characterization of MgO in terms of environmental production processes, calcination temperatures, reactivity, and micro-physical properties. Relationships between different MgO composites and engineered cementitious composites are then established. Next, the influence of MgO incorporation on the properties of cementitious materials and indoor environmental quality are summarized. Finally, future research directions are discussed.

2. Materials and Methods

2.1. Production of Magnesia and Its Use in Cementitious Materials

The construction industry is responsible for a multitude of environmental problems worldwide. It is an energy and resource-intensive industry that generates significant emissions and waste. While steel and concrete are the most commonly used construction materials, each has its own advantages and disadvantages based on price, properties, and structural capabilities. Recently, however, there has been increased concern about the environmental impact of their production and use. Cement, for example, is widely known to be a key ingredient in the production of concrete used in the construction of buildings and other physical infrastructures. The production of cement, in turn, consumes a significant amount of energy. Despite significant technological advances, the world continues to be plagued by health risks and other environmental disasters caused by cement manufacturing companies. Emissions from cement manufacturing harm the environment, degrade air quality, and have a major impact on climate change, contributing significantly to global warming [17].

For avoiding environmental problems, scientists are always searching for sustainable structural materials that show high performance during both processing and application. One way to overcome this issue is to partially replace Portland cement with industrial waste products, e.g., blast furnace slag, fly ash, micro-silica, natural pozzolans, and limestone fillers. Supplementary cementitious materials (SCMs) contribute to the qualities of hardened concrete by hydraulic and/or pozzolanic activity when used in conjunction with Portland cement, Portland limestone cement, or blended types of cement. Supplementary cementitious materials provide long-term and performance benefits to persons who construct and occupy various structures [18]. Several performance factors, including improved workability and consolidation, flexural and compressive strengths, pumpability, resistance to chlorides and sulfates, lower temperatures for mass concrete, mitigation of alkali-silica reaction, and decreased permeability, have contributed to the growing use of these environmentally friendly materials. The use of cementitious blends not only results in stronger, more lasting high-performance concretes but also helps minimize the global climate impact by cutting energy consumption and greenhouse gas emissions. These materials provide a substantial contribution to environmentally friendly buildings. The use of these materials in concrete manufacturing consumes less energy and delivers increased efficiency and building performance [19,20]. These SCMs are not only effective in lowering the environmental impact but also have the potential to enhance the durability of concrete. Some SCMs can modify microstructure by forming additional hydration products. This refines the pore structure and decreases the penetrability of concrete. Magnesia (MgO)-based cementitious composites are another foremost approach towards sustainable design and for promising to attain targets. It is possible to produce environmentally friendly cement with a high MgO content associated with reduced CO_2 emissions.

Cement filler substitutions alter microstructural development in a variety of ways. The particle effect (on hydrate nucleation and dilution of the reactant in a larger volume of water) is distinguished from the hydration of the fillers in the cementitious matrix. As already stated, siliceous SCMs provide silica (reacting with calcium aluminate hydrates to form a new stable) phase-filling space during the hydration and curing. In the second part of the dissertation, we aim to better understand the formation of cement paste and mortar, such as stratlingite, and their influence on the space-filling properties of mortars [21].

Magnesium, at 2.3 percent by weight, is the eighth most common metal in the Earth's crust and is found in a variety of volcanic rocks like olivine, magnesite, and iron oxide. Magnesium is, indeed, the third most common compound in ocean water, with amounts of around 1300 parts per million. The MgO demand is currently 14 million tons per year (USGS, 2012), in comparison to over 2.6 billion tons for OPC, with current prices of about GBP 200 per ton for responsive MgO (calcined), relative to BGP 70 per ton for OPC. The fresh method is used in most industries to produce cement, and it comprises of the following steps: refining and heterogenization of raw materials (to collect raw flour); clink erization of the fresh flour in domestic fuels (to produce clinker); resulting in clinker cooling; refining of clinker and application of gypsum to produce cement; packing and shipment of the end product. This method consumes a lot of energy and produces a lot of air pollution because it needs temperatures as high as 1400 degrees Celsius. Magnesia (magnesium oxide, MgO) is made mostly by calcining magnesite, which is usually the method of making lime from limestone. Seawater and brine streams, as well as other sources, provide a smaller proportion of the world's MgO [22].

Since concrete's roles in the community are pretty scarce in life, and its hydration compound brucite [Mg (OH)$_2$] appears in only a few commercially feasible geological formations, commercially extracted magnesium oxide (commonly referred to as magnesia or periclase) is not mined specifically. Alternatively, MgO is typically obtained via a dried route from calcination of extracted magnesite mines ($MgCO_3$) or a moist route from magnesium-bearing drilling fluids or ocean water substances. As well as the high energy processing needs via the wet path, calcination of magnesite accounts for the majority of MgO global production [22]. The dried path for MgO processing usually necessitates magnesite crushing prior to calcination via the process, and MgO-alkaline oxide plays an electron donor role in water, as shown in the equations below:

$$MgO_{(s)} + H_2O \rightarrow Mg(OH)^+_{(surface)} + OH^-_{aq} \tag{1}$$

OH$^-$ anions are adsorbed on the positively charged surface:

$$Mg(OH)^+{}_{(surface)} + OH^-{}_{aq} \rightarrow Mg(OH)^+ \cdot OH^-{}_{surface} \tag{2}$$

OH$^-$ anions desorbed from the surface, releasing Mg^{2+} and OH$^-$ ions into the solution:

$$Mg(OH)^+ \cdot OH^-{}_{surface} \rightarrow Mg^{2+} + 2\,OH^-{}_{aq} \tag{3}$$

Ion concentration reaches the solution supersaturation (pH~10.5), at which point the hydroxide starts to precipitate as brucite on the oxide surface:

$$Mg^{2+}{}_{(aq)} + 2\,OH^-{}_{(aq)} \rightarrow Mg(OH)_{(2(s))} \tag{4}$$

Here, it should essentially be aimed at increasing the quantity of CO$_2$ and increase the formation of HHMs, and the general mechanical performance of the formulations obtained by increasing MgO hydration.

Since Fe$_2$O$_3$ and SiO$_2$ contaminants may adversely impact MgO's refractory use, higher-grade MgO necessitates the careful selection of MgCO$_3$-bearing rocks or thermal treatment [22]. The wet path is more complicated chemically, but it usually involves precipitating Mg(OH)$_2$ from a magnesium-rich solution as a way to solve seawater or (more dilute) saltwater. Water is pumped into an MgCl$_2$-rich precipitation and transferred to the groundwater to add pressure in Veendam, the Netherlands. Groundwater has an average magnesium concentration of 1.29–1.35 g/L, which varies by area. As a result, groundwater is a huge source of magnesium. Ion-exchange adhesives can also be used to deborate condensed brines or coastal areas, and sulfate concentrations can be decreased by adding CaCl$_2$ brines to instigate CaSO$_4$•2H$_2$O and yield filtered MgCl$_2$-rich saltwater [22].

2.2. Development of Reactive Magnesia Cements

Increased populations directly reflect improvements in health and mortality rates over time, leading to further population expansion. Rising populations, on the other hand, indicate an increase in pressure on existing social facilities, such as housing. As the demand for housing increases exponentially, the construction sector and production of traditional materials such as cement, steel, aluminum, and wood, will be even more strained. According to studies, the production of traditional building materials, such as cement, consumes a significant amount of thermal and electrical energy resulting in higher construction costs [23].

However, as some have noted, the housing supply is inadequate for a variety of reasons. First, poor urban planning limits urban expansion due to a lack of land and infrastructure. Second, insecure land tenure and high urban land costs are exacerbated by various land tenure regimes and ineffective land administration and governance institutions. Third, since housing finance markets in Africa are underdeveloped, most Africans have to rely on self-financing and incremental construction methods to obtain houses. Most importantly, the high cost of construction puts houses out of reach for the majority of low- and middle-income families [17].

Moreover, such manufacturing processes have a larger carbon footprint and pollute the air, land, and water. For example, studies show that the calcination process used to make cement requires temperatures up to 1450 °C and emits about 0.85 tons of CO$_2$ per ton of cement produced. According to another study, buildings in France account for about 23.5 percent of greenhouse gas pollution due to the use of traditional building materials. In a similar vein, others have claimed that the construction industry is currently unsustainable. These findings suggest that further scientific research is needed to develop building materials that are not only more environmentally friendly and sustainable but also more economical without compromising construction quality [23].

The calcination process of reactive MgO requires a lower temperature (700–1000 °C for reactive MgO vs. 1450 °C for PC), which allows the use of alternative fuels with low calorific values (e.g., refuse-derived fuel and hybrid). By interacting with H_2O and CO_2 to bind CO_2 and build strength, the reactive MgO creates the binding property. Figure 1 shows the variables that affect the hydration of MgO.

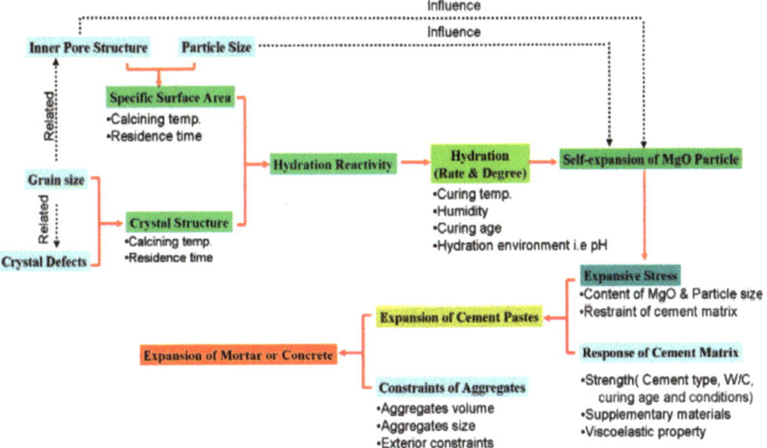

Figure 1. Factors influencing the hydration process [14].

There has already been a revival of excitement in dynamic magnesia (MgO) cements with a high MgO component in recent decades, but most of the research was already published in the field of quality management or web outlets rather than peer-reviewed journals [24]. Whenever the responsive MgO is generated in a carbon-recycling cement kiln, the subsequent CO_2 absorption (by the cement in operation) is taken into account. Based on a 2001 application, Harrison of the Australian company TecEco was granted a patent in the United States that explains the use of expanded curing periods and occasionally steam-curing to manufacture solid blocks made of MgO, pozzolan, and PC [25,26]. The MgO used is made by calcining $MgCO_3$ at a low temperature of 800 °C; this causes layer strain and permeability in MgO samples that would otherwise be coated at higher temperatures. This allows for the precise regulation of MgO reactivity based on treatment conditions and particle shape, ensuring that it hydrates at the same period as the other cementitious materials [25]. Pertinently, this MgO responds much faster than kilned MgO (low-reactivity MgO calcined at >1500 °C), including the free MgO in Portland (clinker), which has been fired, often at extreme temps, and thus normally hydrate slowly, causing cracking within traditional cements as an expansive reaction rate is caused centrally within a hardened substrate [26]. MgO has little impact on the formation of PC hydrate processes after up to four weeks of hydration [27–29]. Many tests have shown that some responsive MgO–PC-blended cements do not absorb a detectable amount of CO_2 from the atmosphere within the period of curing and are hence impossible to be carbon-negative, or even carbon-neutral, in the appropriately limited period. Once MgO is applied to a PC-based device, Cwirzen and Habermehl-Cwirzen [25] found that now the freeze-thaw tolerance, flexural and compressive strengths are decreased due to higher capillary permeability. Figure 2 shows low magnification scanning electron micrographs of fractured surfaces of all types of samples at 14 days.

Nevertheless, when accelerated-carbonation healing criteria were applied to responsive MgO structures, a very opposite result was obtained. Until being split onto 5-mm-thick specimens, MgO/PC/FA- and MgO/FA-based cements became air preserved for two weeks at 98% moisture content (MC). Such specimens were again preserved for another two weeks, whether in the air at 98% MC as a monitor or in monitored CO_2 atmospheres at atmospheric pressures including 5 or 20% CO_2 by volume, at 65% or 95% MC [29].

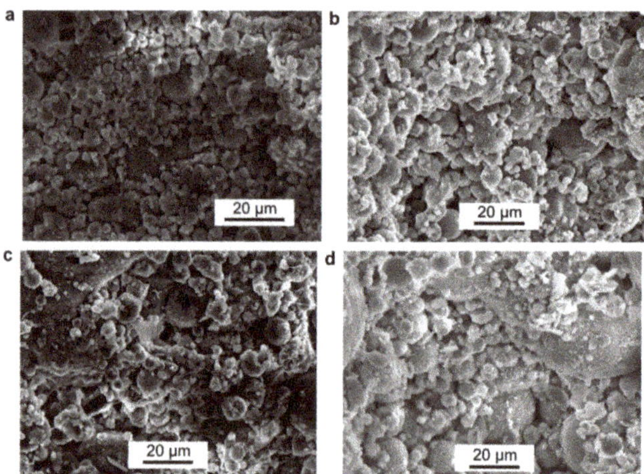

Figure 2. Low magnification secondary electron micrographs after 14 days of curing of the 50% and 90% pfa content mixes: (**a**) $MgO_{0.1}$-$pfa_{0.9}$, (**b**) $(MgO_{0.8}PC_{0.2})_{0.1}$-$pfa_{0.9}$, (**c**) $(MgO_{0.5}PC_{0.5})_{0.1}$-$pfa_{0.9}$, (**d**) $PC_{0.1}$-$pfa_{0.9}$, [24].

Mo and Panesar recently published research on responsive MgO, focusing on the rapid carbonation of MgO/PC blends both with and without the inclusion of surface granulated blast-furnace slag (BFS) [30]. Such cements produced up to 40% MgO, with $MgCO_3 \bullet 3H_2O$ and $CaCO_3$ even as the main carbonate compounds produced (both calcite and aragonite polymorphs). Cement materials were vacuum-dried to extract the capillary humidity before even being subjected to a 99.9% CO_2, 98% MC, allowing for accelerated carbonation of the collections. The existence of MgO was said to change calcite composition, leading to the formation of magnesian calcite that, in combination with the accumulation of $MgCO_3 \bullet 3H_2O$, decreased sample porous structure, densified the microstructure, and increased microhardness [31]. Due to the comparatively harsh carbonation circumstances, it is unclear if this carbonation system will be used commercially or on a broader scale.

2.3. Expansive MgO Cements

It's worth noting that responsive MgO cements are different from limited propor-tion reactive MgO as a comprehensive additive in cement binders, which are commonly used in dam building and other major construction projects, especially in China. This is to substitute for PC's natural hydration shrinkage, which can last weeks or months in operation [32]. The usage of decent low cements or the intensive need for supplemental cementitious materials will help only with the cooling shrinkage of cement paste during toughening, which can be mitigated by using massive cement/concrete edifices. This really is attributable to the fact that cement hydration is strongly exothermic, releasing upwards of 500 J/g of cement. When the temperature goes up in reach of 50 °C, after the concrete has been cured (up to six months after casting), hydration of the cement in such massive amounts of concrete occurs [9,33].

In 1867, Sorel cement or magnesium oxychloride cement (MOC) was discovered. MOC was prepared by mixing magnesium oxide (MgO) with magnesium chloride ($MgCl_2$) [33]. The $MgO/MgCl_2$ and $H_2O/MgCl_2$ molar ratios are the main parameters, which potentially affect the mechanical properties of MOC [34]. The main hydration products of MOC, which are responsible for its hardening and strength, are $5Mg(OH)_2 \bullet MgCl_2 \bullet 8H_2O$ (phase 5), $3Mg(OH)_2 \bullet MgCl_2 \bullet 8H_2O$ (phase 3) and $2 Mg(OH)_2 \bullet MgCl_2 \bullet 8H_2O$ (phase 2). The com-position of hydrate phases mainly depends on the $MgO/MgCl_2$ molar ratio [35]. The mechanism of hydrate phase formation includes three steps: the first is the neutralizing process in which MgO powder is neutralized by free H^+ produced from the dissociation of

MgCl$_2$ crystals in water. The second includes the formation of bi-nuclear, tri-nuclear, and poly-nuclear complexes {Mg$_x$(OH)$_y$(H$_2$O)$_z$}$^{2x-y}$ by the hydrolyzing-bridging reaction. In the final step, the condensation of these phases and the adsorption of Cl$^-$ (to equalize the positive charge on complexes) have occurred, leading to the formation of an amorphous gel, which crystallized in a few days or weeks [35]. MOC characterizes by low thermal conductivity, high early strength, high firing, and good abrasion resistivity [36]. Although advantageous properties, the MOC showed poor water resistivity, limiting its application for practical engineering projects.

Based on the mechanism of MOC formation [35], MgO cement has been considered a good alternative to traditional Portland cement. MgO-based composites have been characterized by their high strength, early hardening, and strong adhesion strength. However, the inherent brittleness of these composites may restrict the number of application areas in practice. To overcome this issue and extend the application range of these composites within the construction industry, MOC-based engineered cementitious composites (MOC-ECC) have been developed.

When the concrete cools down, it expands, resulting in a crack-prone dam. Various structural engineering ventures have long used shrinkage-compensating and expansive cements [36–39]. These really are traditionally dependent on applying ye'elimite (Ca$_4$Al$_6$O$_{12}$SO$_4$) with anhydrite (CaSO$_4$) to cements to increase aluminate and sulfate concentrations, resulting in extensive value [Ca$_6$Al$_2$(SO$_4$)$_3$(OH)$_{12}$•26H$_2$O] crystals on such hydration [40]. Traditional shrinkage-compensating cements, on the other hand, are inappropriate for massive structural parts where shrinkage is often detected as a product of cooling after a preliminary exergonic hydration reaction instead of autogenous or dried shrinkage of the cement hydrates—as shrinkage occurs far after the intended expansive materials have developed. MgO-extensive cements have been gaining popularity for this reason. The extensive hydration of MgO to Mg(OH)$_2$, which results in a 117 percent molar solid mass transfer, is used in these studies [41–43]. One study was conducted to investigate the mechanical and morphological properties of carbonized corn stalk used to reinforce polyester composites in the manufacture of environmentally friendly composites [32]. A comparison of the results reveals two important findings. For starters, agro-waste materials could be employed in their natural state in reinforcing applications, such as bamboo in cementitious applications. Second, the agro-waste materials might be treated or employed as chemical admixtures in reinforcing biocomposites, implying that they needed to be treated before being used in reinforcement applications [32,44,45].

The sustainability advantages of MgO involve (i) adequacy of carbonate to gain vigor/strength in relation to this, (ii) appreciable durability increase due to the higher resistance of the hydration and carbonation products in assailant environments, (iii) lower susceptibility to smudginess enabling the utilization of considerable amounts of waste and industrial by-products, and (iv) probable entirely recyclable where MgO is used alone as the binder as its carbonation time course produces magnesium carbonates, which are the dominant resource for the production of magnesia. Interchangeably, accurate restrictions come into being concerning the production and application of MgO cements in the construction sector [43].

2.4. Recent Developments for Building Design

A common thread running through these research studies is that they all aim to solve two major problems. The first is to reduce the impact of the construction industry on climate change by promoting the use of alternative materials. The implication of the two factors (depletion of non-renewable resources, high pollution levels) makes it necessary to refocus on the need for sustainability in the construction industry. On the one hand, it is necessary to ensure that the raw materials used in construction, such as cement and sand, are not used up, but on the other hand, it is also necessary to ensure that the results of the construction industry (buildings and infrastructure) do not emit significant amounts of carbon dioxide. As a result, several efforts, as well as countless research studies, have been developed over the years to ensure the sustainability of the construction sector [32,44].

Another industrial hurdle is the pre-processing stage that some agricultural wastes must undergo before they can be combined with conventional materials. In one study, wheat and barley straw fibers were treated with boiling water and linseed oil to reduce water absorption while improving binder compatibility and adhesion; they were used to produce lightweight composites for building insulation. In another study, alkali was used to treat agricultural waste in the development of composite materials made of rice straw, magnesium cement adhesive, and a foaming agent (NaOH) [32]. Similarly, researchers were observed burning other solid wastes such as peanut, rice, and barley husks to produce ash that could be mixed with traditional materials for the construction of bricks and masonry components. It was also found that the ash had to be further sieved before being incorporated into the bricks [44].

It is very well known that the incorporation of a low amount of short fibers (Figure 3) into the cementitious matrix is a very effective solution for preventing brittleness and improving the tensile ductility of PC-based composite materials [45]. Engineered cementitious composites (ECC), which adopt polymeric fibers at a typical 2% by volume mixture, are a good example of effective and successful fiber supplements. ECC shows strain-hardening behavior like metal, and thus, the tensile stress of these types of composites continue to increase even within the presence of cracks. Cracking is considered one of the most common forms of deterioration in concrete structures leading to strength loss, thermal discomfort, and energy consumption. Cracks in buildings are inevitable and can be created at nearly every phase of the material's service life by thermal gradients, over-loading, or chemical attacks.

The high cement content is required to produce ECC mixes for providing strain hardening behavior and reducing the matrix toughness. These characteristics of ECCs offer an attractive change for utilization of reactive MgO cement combined with CO_2 curing and has encouraged the development of a novel version of ECC built upon the MgO-fly ash-CO_2 system. This innovative MgO-ECC has a tensile strain ductility of more than 5% and successfully sequesters approximately 30% CO_2 by mass of MgO within 24 h, which, in turn, provides new sustainable building design applications for the carbonated MgO cement [44–47].

Wu et al. [43,44] investigated the cracking behavior of concrete made with reactive MgO and flew ash cured with an accelerated carbonating process for one, three and seven days. The study revealed that the carbonation curing densifies the interfacial bonding system, resulting in a significant increase of the retarded tensile strength at first cracking, which in turn, has considerable influence on the fracture properties of concrete. The recently developed ECC-supported reactive MgO-fly ash blends show a guarantee in building up self-healing ability. The typical width of the multiple micro-cracks formed under uniaxial tension measured but 60 μm. The ECC's matrix contains a high volume of reactive MgO, which is not completely responded to during the accelerated CO_2 carbonation. The unreacted MgO and its hydration products create the potential for subsequent dissolution and precipitation in microcracks that will facilitate the self-healing process [46–48].

Reactive MgO is not only used for self-healing but also frequently used to optimize the shrinkage behavior of concrete. While M92-200 is a moderately reactive grade of MgO that has been used for a variety of applications, researchers have also used highly reactive grades of MgO to reduce shrinkage in concrete [49]. Both reactive MgO types, therefore, have a high potential for self-healing of drying shrinkage cracks. The development of internal stresses within the cement matrix is caused by the expansion of MgO in the early stages of hydration. The reason for equalizing the shrinkage stress in cement is well known. The optimum proportions of MgO in PC lead to suitable expansion by densification of the microstructure through partial MgO hydration.

a: role of short fibers during the
micro cracking process

b: role of long fibers during the
macrocracking process

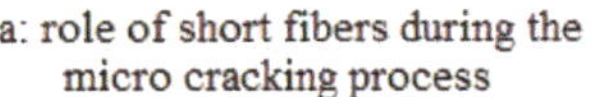

1: rupture of the concrete without fibers
2: rupture of the concrete with short fibers

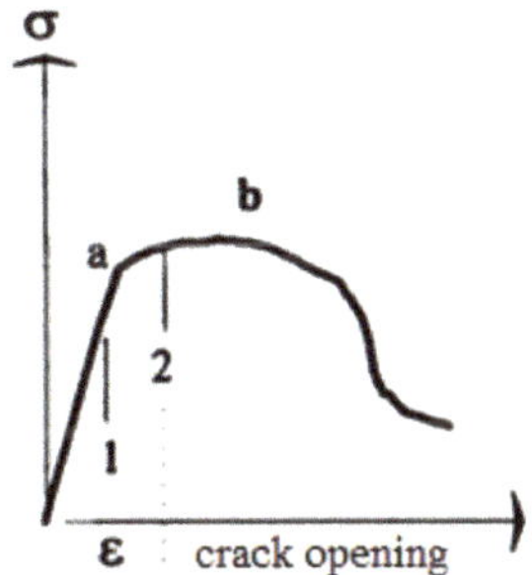

Figure 3. Structure of short and long fibers controlling microcracks and their influence on the stress–crack opening curve [48].

The conversion of lignin to high value-added products plays a key role in the economic viability of a biorefinery. For example, some potential new applications for lignin are the production of nano-scale structures for metal absorption captured in water or air pollutant; generation of inexpensive-effective pyrolysis processes to obtain simple compounds as industrial solvents to replace petroleum-derived compounds such as toluene, xylene, and benzene; or development of 3D printing materials designed for biomedical applications (artificial tissues as support for antioxidant, antimicrobial, or biodegradable compounds from lignin). Optimizing the catalytic mechanisms using lignin as raw material will facilitate the development of improved sustainable materials [46].

In a concert of emerging strategies countering such deterioration, the self healing of concrete cracks has been progressively contemplated. Although the addition of fiber was shown to have a positive effect in reducing water permeability, increasing compressive and flexural strengths of matrixes, controlling micro-cracks, and improving impact resistance— the improvement was far from ideal to solve the mechanical and durable issues that MOC encounters in practical applications [46,48].

2.5. Life Cycle Assessment for Building Design

The demand for cement is constantly increasing, due to the growth of the world population, making it the most used building material, reaching a production of 10 billion tons per year. Because of these huge quantities, the impact on the environment is substantial in terms of embodied energy consumption, raw materials required, and greenhouse gas emissions. Indeed, the latter aspect amounts to around 5–7% of the anthropogenic carbon dioxide emitted, contributing to global warming, mostly because of the Portland cement, one of the widely used binders of modern concrete mixtures, which is not environmentally friendly [49].

A building's total life cycle energy consumption is divided into two categories: embodied energy and operational energy. Embodied energy is the entire quantity of non-renewable primary energy necessary for all direct and indirect processes associated with the construction of a building, maintenance, and end-of-life, whereas operational energy is the energy used during the building's use stage [50]. To reduce carbon dioxide emissions, the architectural industry has concentrated on reducing the usage of fossil fuels throughout the construction stage. However, in order to achieve the net-zero policy target, efforts must be made to decrease the embodied energy that is generated during the collection and production of building materials. Nevertheless, even in the case of green and sustainable architecture, which basically aims to minimize carbon dioxide emissions and environmental impacts over the life cycle, building foundations or structural frames are mostly made of concrete, which requires a lot of energy during the manufacturing process. This is because of the difficulties in developing viable alternatives to the current concrete-based strategy when numerous considerations like structural safety and economic efficiency are considered. As a result, developing a new binder that is more environmentally friendly and reduces the embodied energy of concrete could be a useful alternative to finding a construction method that substitutes concrete foundations [51].

There are over 100,000 materials in our world, and suitable material selection is critical. Different factors can be considered to select alternative materials, depending on the required functional properties and the final cost. Today, more attention must be devoted to sustainability. Sustainable materials can be defined as materials derived from renewable resources. They must have a zero/minimal impact on the environment and society for their extraction and production. Examples are recycled metals, bio-based polymers, and materials for renewable energy. In some cases, biomaterials can be interesting for combined applications, such as bioremediation and fuel production [49]. There are also eco-friendly binders that lead to less carbon dioxide emissions than OPC, namely including magnesium phosphate cement, geopolymer, alkali-activated slag cement, and super sulfate cement [51–60].

Sinka et al. [61] developed a bio-based wall panel consisting of MgO cement (as a binder) and hemp shives and compared its CO_2 emission performance to the other wall panels made with traditional materials. Figure 4 illustrates that the emissions of a wooden frame wall, which is filled with the mineral wool, emits only 16.1 kg/CO_2 eq m^2, 27% more CO_2 of the magnesium-hemp panel, as both wall types consist of a load-bearing wooden frame, the largest part of emissions (around 50%) comes from the mineral wool. More interestingly, the magnesium-hemp material produces significantly fewer emissions compared to the traditionally used materials because hemp shives could absorb the CO_2 that was built in the wall material.

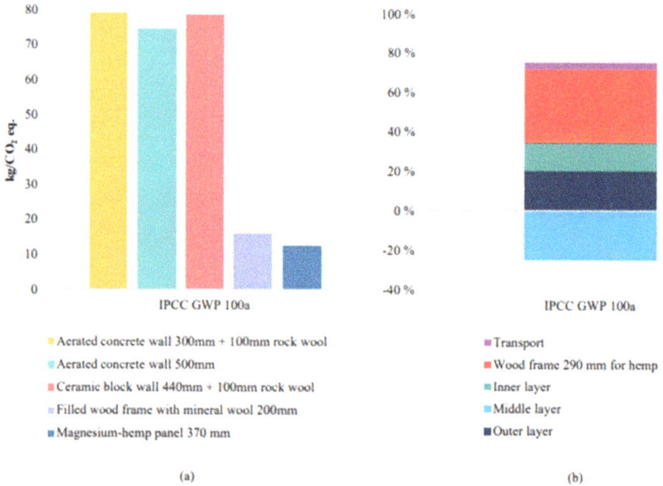

Figure 4. (**a**) Comparison of different wall types and their greenhouse grass emissions; (**b**) emissions by percentage of hemp-magnesium panel parts [50].

A recent study conducted by Li et al. [60] investigated a life cycle assessment (LCA) of an innovative and sustainable magnesium oxide structural insulated sandwich panel (MgO SIP) used for a high-performance smart home in Vancouver. For those purposes, the authors constructed a prototype house (Figure 5) and compared the environmental impacts across six parameters for the MgO SIPs, traditional SIPs, and traditional stick-frame construction across the life cycle phases of raw material extraction, manufacturing, transportation, construction, and operation. The findings of this study indicated that the MgO SIPs do not outperform conventional alternatives, notably because of the long-distance transportation of materials. However, further LCA of hypothetical scenarios shows that MgO SIPs have a great potential to become more environmentally friendly than the conventional alternatives by sourcing MgO domestically, implementing onsite manufacturing, and eliminating oriented strand boards.

Figure 5. The prototype structure for the study [50].

3. Conclusions

Research on MgO and MgO-based cementitious composites is very important nowadays when most of the CO_2 emission and heat consumption is made for heating and cooling purposes in buildings. MgO-based building elements (for example, wall panels, roof decking) can effectively be used for energy harvesting in buildings and provide a comfortable temperature inside the building. This study refers to the characteristics and developments of MgO-based composites.

- MgO-based composite application in cementitious materials and energy-efficient buildings, and summaries of the numerical and experimental studies on these materials, have shown that MgO cement-based composites can play an important role in terms of sustainable and environmentally friendly building design.
- It has been emphasized in studies that MgO-based composites can provide increases in temperature comfort in buildings.
- It is very important that MgO, which has been recently used in buildings with a wide variety of methods, is suitable for climatic conditions, and the type of application should be selected according to the climate.
- It's crucial to form an oxidation and reduction method that can remove individual carbon stratum and isolate them without changing their composition until magnesium oxide can be used as an intermediate in the development of a monolayer or few-layer MgO sandwich panel parts.

- It has been observed that although chemical reduction of magnesium oxide is generally thought to be the best process for mass production of MgO composites, scientists have struggled to complete the challenge of manufacturing lightweight and fiber-reinforced MgO sheets of the same quality as mechanical exfoliation, but on a much larger scale.
- We should expect magnesium to become much more commonly used in consumer and industrial applications until this problem is resolved.
- Finally, MgO-based engineered cementitious modified sustainable building materials provide crack toughening mechanisms improving the bridging stress of the cementitious matrix and make the micro-structure much denser, which, in turn, increases the load-carrying capacity of the composite under mechanical stresses.

Author Contributions: Conceptualization, S.K., S.E. and R.A.G.L.; methodology, S.K., S.E.; formal analysis, S.K., S.E., R.A.G.L.; investigation, S.K.; resources, S.K., S.E. and R.A.G.L.; data curation, S.K., S.E.; writing—original draft preparation, S.K., S.E.; writing—review and editing, S.K., S.E. and R.A.G.L.; visualization, S.K., S.E. and R.A.G.L.; supervision, R.A.G.L.; project administration, S.K., S.E., R.A.G.L.; funding acquisition, R.A.G.L. All authors have read and agreed to the published version of the manuscript.

Funding: The authors wish to thank CEU San Pablo University Foundation for the funds dedicated to the Project Ref. USP CEU-CP20V12 provided by CEU San Pablo University.

Institutional Review Board Statement: Not applicable.

Informed Consent Statement: Not applicable.

Data Availability Statement: The data presented in this study are available on request from the corresponding author.

Conflicts of Interest: The authors declare no conflict of interest.

References

1. Freedonia Group. *World Construction Aggregates- Demand and Sales Forecasts, Market Industry Study No.3389*; Freedonia Group: Cleveland, OH, USA, 2016; p. 390.
2. USGS. *Commodity Statistics and Information Mineral. Yearbooks*; USA Geological Survey: Washington, DC, USA, 2015.
3. Marinković, S.; Radonjanin, V.; Malesev, M.; Ignjatovic, I. Comparative environmental assessment of natural and recycled aggregate concrete. *Waste Manag.* **2010**, *30*, 2255–2264. [CrossRef]
4. De Schepper, M.; Heede, P.; Driessche, I.; de Belie, N. Life cycle assessment of completely recyclable concrete. *Materials* **2014**, *7*, 6010–6027. [CrossRef] [PubMed]
5. Kurda, R.; de Brito, J.; Silvestre, J.D. Combined economic and mechanical performance optimization of recycled aggregate concrete with high volume of fly ash. *Appl. Sci.* **2018**, *8*, 1189. [CrossRef]
6. Kurda, R.; de Brito, J.; Silvestre, J.D. Water absorption and electrical resistivity of concrete with recycled concrete aggregates and fly ash. *Cem. Concr. Compos.* **2019**, *95*, 169–182. [CrossRef]
7. Berndt, M.L. Properties of sustainable concrete containing fly ash, slag and recycled concrete aggregate. *Constr. Build. Mater.* **2009**, *23*, 2606–2613. [CrossRef]
8. Kou, S.C.; Poon, C.S.; Agrela, F. Comparisons of natural and recycled aggregate concretes prepared with the addition of di erent mineral admixtures. *Cem. Concr. Compos.* **2011**, *33*, 788–795. [CrossRef]
9. Ferdous, W.; Manalo, A.; Wong, H.; Abousnina, R.; Ajarmeh, O.; Zhuge, Y.; Schubel, P. Optimal design for epoxy polymer concrete based on mechanical properties and durability aspects. *Constr. Build. Mater.* **2020**, *232*, 117–229. [CrossRef]
10. Canterford, J.H. Magnesia—An important industrial mineral: A review of processing options and uses Miner. *Process. Extr. Metall. Rev.* **1985**, *2*, 57–104. [CrossRef]
11. Eubank, W.R. Calcination studies of magnesium oxides. *J. Am. Ceram. Soc.* **1951**, *34*, 225–229. [CrossRef]
12. Wright, J.M.; Colling, A. *Seawater: Its Composition, Properties and Behaviour*, 2nd ed.; Elsevier Science: Oxford, UK, 1995.
13. Seeger, M.; Otto, W.; Flick, W.; Bickelhaupt, F.; Akkerman, O.S. Magnesium compounds. In *Ullmann's Encyclopedia of Industrial Chemistry*; Wiley-VCH Verlag GmbH & Co. KGaA: Weinheim, Germany, 2000.
14. Mo, L.; Deng, M.; Tang, M.; Al-Tabbaa, A. MgO expansive cement and concrete in China: Past, present and future. *Cem. Concr. Res.* **2014**, *57*, 1–12. [CrossRef]
15. Venkatarama Reddy, B.V. Sustainable materials for low carbon buildings. *Int. J. Low Carbon Technol.* **2009**, *4*, 175–181. [CrossRef]
16. Vance, K.; Falzone, G.; Pignatelli, I.; Bauchy, M.; Balonis, M.; Sant, G. Direct Carbonation of $Ca(OH)_2$ Using Liquid and Supercritical CO_2: Implications for Carbon-Neutral Cementation. *Ind. Eng. Chem. Res.* **2015**, *54*, 8908–8918. [CrossRef]

17. Bontempi, E.; Sorrentino, G.P.; Zanoletti, A.; Alessandri, I.; Depero, L.E.; Caneschi, A. Sustainable materials and their contribution to the sustainable development goals (SDGs): A critical review based on an Italian example. *Molecules* **2021**, *26*, 1407. [CrossRef]

18. Snellings, R.; Mertens, G.; Elsen, J. Supplementary Cementitious Materials. *Rev. Mineral. Geochem.* **2012**, *74*, 211–278. [CrossRef]

19. Lang, E. Blast furnace cements. In *Structure and Performance of Cements*, 2nd ed.; Bensted, J., Barnes, P., Eds.; Spon Press: London, UK, 2002; pp. 310–325.

20. Sonebi, M.; Ammar, Y.; Diederich, P. Sustainability of cement, concrete and cement replacement materials in construction. In *Sustainability of Construction Materials*; Woodhead Publishing: Oxford, UK, 2016; pp. 371–396.

21. Gosselin, C.; Scrivener, K. *Microstructural Development of Calcium Aluminate Cement-Based Systems with and without Supplementary Cementitious Materials*; EPFL: Lausanne, Switzerland, 2009; p. 234.

22. Walling, S.A.; Provis, J.L. Magnesia-Based Cements: A Journey of 150 Years, and Cements for the Future? *Chem. Rev.* **2016**, *116*, 4170–4204. [CrossRef]

23. Sodangi, M.; Kazmi, Z.A. Integrated evaluation of the impediments to the adoption of coconut palm wood as a sustainable material for building construction. *Sustainability* **2020**, *12*, 7676. [CrossRef]

24. Vandeperre, L.J.; Liska, M.; Al-Tabbaa, A. Microstructures of reactive magnesia cement blends. *Cem. Concr. Compos.* **2008**, *30*, 706–714. [CrossRef]

25. Cwirzen, A.; Habermehl-Cwirzen, K. Effects of reactive magnesia on microstructure and frost durability of Portland cement–based binders. *J. Mater. Civ. Eng.* **2013**, *25*, 1941–1950. [CrossRef]

26. Jackson, P.J. Portland cement: Classification and manufacture. In *Lea's Chemistry of Cement and Concrete*; Hewlett, P.C., Ed.; Butterworth-Heinemann: Oxford, UK, 2003.

27. Vandeperre, L.J.; Liska, M.; Al-Tabbaa, A. Hydration and mechanical properties of magnesia, pulverized fuel ash, and Portland cement blends. *J. Mater. Civ. Eng.* **2008**, *20*, 375–383. [CrossRef]

28. Harrison, J. New cements based on the addition of reactive magnesia to portland cement with or without added pozzolan. In Proceedings of the CIA Conference: Concrete in the Third Millennium, CIA, Brisbane, Australia, 17–19 July 2003.

29. Vandeperre, L.J.; Al-Tabbaa, A. Accelerated carbonation of reactive MgO cements. *Adv. Cem. Res.* **2007**, *19*, 67–79. [CrossRef]

30. Mo, L.; Panesar, D.K. Effects of accelerated carbonation on the microstructure of Portland cement pastes containing reactive MgO. *Cem. Concr. Res.* **2012**, *42*, 769–777. [CrossRef]

31. Maraveas, C. Production of sustainable construction materials using agro-wastes. *Materials* **2020**, *13*, 262. [CrossRef] [PubMed]

32. Liu, Z.; Wang, S.; Huang, J.; Wei, Z.; Guan, B.; Fang, J. Experimental investigation on properties and microstructure of magnesium oxychloride cement preparedwith caustic magnesite and dolomite. *Constr. Build. Mater.* **2015**, *85*, 247–255. [CrossRef]

33. Li, Z.; Chau, C.K. Influence of molar ratios on properties of magnesium oxychloride cement. *Cem. Concr. Res.* **2007**, *37*, 866–870. [CrossRef]

34. Abdel-Gawwad, H.A.; Khalil, K.A. Preparation and characterization of one-part magnesium oxychloride cement. *Constr. Build. Mater.* **2018**, *189*, 745–750. [CrossRef]

35. Bensted, J.; Barnes, P. *Structure and Performance of Cements*, 2nd ed.; Spon Press: London, UK, 2017.

36. Du, C. A review of magnesium oxide in concrete. *Concr. Int.* **2005**, *27*, 45–50.

37. Bamforth, P.B. In situ measurement of the effect of partial Portland cement replacement using either fly ash or ground granulated blast-furnace slag on the performance of mass concrete. *Proc. Inst. Civ. Eng.* **1980**, *69*, 777–800. [CrossRef]

38. Bensted, J. Gypsum in cements. In *Structure and Performance of Cements*, 2nd ed.; Bensted, J., Barnes, P., Eds.; Spon Press: London, UK, 2002.

39. Nagataki, S.; Gomi, H. Expansive admixtures (mainly ettringite). *Cem. Concr. Compos.* **1998**, *20*, 163–170. [CrossRef]

40. Chatterji, S. Mechanism of expansion of concrete due to the presence of dead-burnt CaO and MgO. *Cem. Concr. Res.* **1995**, *25*, 51–56. [CrossRef]

41. Unluer, C.; Al-Tabbaa, A. Impact of hydrated magnesium carbonate additives on the carbonation of reactive MgO cements. *Cem. Concr. Res.* **2013**, *54*, 87–97. [CrossRef]

42. Unluer, C.; Al-Tabbaa, A. Characterization of light and heavy hydrated magnesium carbonates using thermal analysis. *J. Therm. Anal. Calorim.* **2014**, *115*, 595–607. [CrossRef]

43. Li, V.C.; Mishra, D.; Wu, H.-C. Matrix design for pseudo-strain-hardening fibre reinforced cementitious composites. *Mater. Struct.* **1995**, *28*, 586–595. [CrossRef]

44. Li, V.C.; Wang, S.; Wu, C. Tensile strain-hardening behavior of polyvinyl alcohol engineered cementitious composite (PVA-ECC). *Mater. J.* **2012**, *98*, 483–492.

45. Mo, L.; Panesar, D.K. Accelerated carbonation—A potential approach to sequester CO_2 in cement paste containing slag and reactive MgO. *Cem. Concr. Compos.* **2013**, *43*, 69–77. [CrossRef]

46. Vásquez-Garay, F.; Carrillo-Varela, I.; Vidal, C.; Reyes-Contreras, P.; Faccini, M.; Mendonça, R.T. A review on the lignin biopolymer and its integration in the elaboration of sustainable materials. *Sustainability* **2021**, *13*, 2697. [CrossRef]

47. Siddique, R.; Naik, T.R. Properties of concrete containing scrap tire rubber-an overview. *Waste Manag.* **2004**, *24*, 563–569. [CrossRef]

48. Wang, Y.; Wei, L.; Yu, J.; Yu, K. Mechanical properties of high ductile magnesium oxychloride cement-based composites after water soaking. *Cem. Concr. Compos.* **2019**, *97*, 248–258. [CrossRef]

49. Forero, J.A.; Bravo, M.; Pacheco, J.; de Brito, J.; Evangelista, L. Fracture Behaviour of Concrete with Reactive Magnesium Oxide as Alternative Binder. *Appl. Sci.* **2021**, *11*, 2891. [CrossRef]
50. Abouhamad, M.; Abu-Hamd, M. Life Cycle Environmental Assessment of Light Steel Framed Buildings with Cement-Based Walls and Floors. *Sustainability* **2020**, *12*, 10686. [CrossRef]
51. Kim, H.-S.; Kim, I.; Yang, W.-H.; Moon, S.-Y.; Lee, J.-Y. Analyzing the Basic Properties and Environmental Footprint Reduction Effects of Highly Sulfated Calcium Silicate Cement. *Sustainability* **2021**, *13*, 7540. [CrossRef]
52. Skalny, J.; Skalny, J.P.; Odler, I. *Materials Science of Concrete: Calcium Hydroxide in Concrete*; Wiley-Blackwell: Hoboken, NJ, USA, 2001.
53. Wagh, A.S. *Chemically Bonded Phosphate Ceramics: Twenty-First Century Materials with Diverse Applications*; Elsevier: Amsterdam, The Netherlands, 2016.
54. Roy, D.M. New strong cement materials: Chemically bonded ceramics. *Science* **1987**, *235*, 651–658. [CrossRef] [PubMed]
55. Provis, J.L.; Van Deventer, J.S.J. *Geopolymers: Structures, Processing, Properties and Industrial Applications*; Elsevier: Amsterdam, The Netherlands, 2009.
56. Davidovits, J. *Geopolymer Chemistry and Applications*, 3rd ed.; Davidovits, J., Ed.; Geopolymer Institute: Saint-Quentin, France, 2011.
57. Roy, D.M. Alkali-activated cements opportunities and challenges. *Cem. Concr. Res.* **1999**, *29*, 249–254. [CrossRef]
58. Shi, C.; Roy, D.; Krivenko, P. *Alkali-Activated Cements and Concretes*; CRC Press: Boca Raton, FL, USA, 2006.
59. Woltron, G. The Utilisation of GGBFS for Advanced Supersulfated. *World Cem. Mag.* **2009**, *10*, 157–162.
60. Li, P.; Froese, T.M.; Cavka, B.T. Life Cycle Assessment of Magnesium Oxide Structural Insulated Panels for a Smart Home in Vancouver. *Energy Build.* **2018**, *175*, 78–86. [CrossRef]
61. Sinka, M.; Korjakins, A.; Bajare, D.; Zimele, Z.; Sahmenko, G. Bio-based construction panels for low carbon development. *Energy Procedia* **2018**, *147*, 220–226. [CrossRef]

 *sustainability*

Article

Purpose Adequacy as a Basis for Sustainable Building Design: A Post-Occupancy Evaluation of Higher Education Classrooms

Vicente López-Chao [1,*] and Vicente López-Pena [2,*]

1 Architectural Graphics Department, School of Architecture, Campus da Zapateira, Universidade da Coruña, 15008 Coruña, Spain
2 Department of Mechanical Engineering and Industrial Design, Faculty of Engineering, Campus de Puerto Real, Universidad de Cádiz, 11003 Cádiz, Spain
* Correspondence: v.lchao@udc.es (V.L.-C.); vicente.lopez@uca.es (V.L.-P.)

Abstract: Building design is one of the essential elements to consider for maximizing the sustainability of construction. Prior studies on energy and resource consumption and on indoor environmental quality indicators (IEQs) are increasingly frequent; however, attention has not been focused on design as supporting the function performed within architecture. Educational buildings have specific conditions related to teaching methodologies, including activating students and promoting participation and interaction in the classroom. This manuscript aims to explore whether the social dimension of physical space in educational settings can explain a student's academic outcome. For this, the Learning Environment and Social Interaction Scale was designed and validated and applied to 796 undergraduate students at the University of Coruña, and multiple linear regression analysis was applied to the academic results. The results display a structure comprising five factors; these include novelties such as the division of conventional IEQs into two groups: the workspace and the classroom environment. In addition, place attachment, the design of the classroom as a facilitator of social interaction, the learning value of social interaction, and the satisfaction of the IEQ demonstrated their influence on the academic result.

Keywords: architecture; building evaluation; functional adequacy; human-centered; IEQ; learning space; place attachment; social interaction; social participation; sustainable building

Citation: López-Chao, V.; López-Pena, V. Purpose Adequacy as a Basis for Sustainable Building Design: A Post-Occupancy Evaluation of Higher Education Classrooms. *Sustainability* 2021, 13, 11181. https://doi.org/10.3390/su132011181

Academic Editor: Roberto Alonso González Lezcano

Received: 14 August 2021
Accepted: 8 October 2021
Published: 11 October 2021

Publisher's Note: MDPI stays neutral with regard to jurisdictional claims in published maps and institutional affiliations.

1. Introduction

Indoor environment quality indicators have been recognized as main features of sustainable design. Therefore, research on their influence is increasingly abundant [1–3]. However, an evident sustainability factor—the suitability or usefulness of the environment for the use of the building—has not received the same attention. It seems logical that if the relationship between built space and its function is consistent, the energy and resources required will be more efficient.

Previous studies have focused on the technical measurement of learning spaces through indoor environment quality indicators, which include lighting, ventilation, thermal levels, connection with nature, acoustics, etc. [4–9]. The validity of this approach is proven and of great relevance to understand to what extent and how the indoor environment can influence the users of the space. The IEQ has become a key factor in the design and construction of buildings, since internal conditions can significantly influence the well-being, productivity, health and safety of people [10]. Therefore, in recent decades, different certifications have been designed, such as LEED (Leadership in Energy and Environment Design), the WELL Building Standard and Fitwel, for different types of buildings. However, these focus on low energy consumption or technical aspects of the building or on the health and comfort of the users. However, both the socio-psychological factors and those related to intended activities are significant in this field.

Since 1960, attempts have been made to study social interactions and the user's perception of the environment, an issue that emerged in the field of environmental psychology through post-occupancy evaluation (POE) studies [11]. Sustainable design is not only about reducing emissions and saving energy but also about providing the necessary comfort in the environment for the development of human activities [12,13]. For this reason, in POE studies, it is usual to include user characteristics, work processes, user satisfaction regarding the possibilities of interaction with their colleagues, and privacy and comfort [14].

Current studies have explored the effects of poor environments on cognitive functions, including social cognition [15]. Advances in this field confirm that subjective issues play an important role in user behavior, such as attitude, social customs, or perceived behavioral control, as well as intentionality [16]. Social values, cultural differences, and personality traits have also become factors to be valued among the scientific community [17–19]. However, there is still some uncertainty about the selection and use of appropriate contextual, social and personal variables; this could be addressed through the implementation of interdisciplinary frameworks [20]. In addition, recent studies have emphasized the need to consider the relationships and interactions between physical or technical variables and personal and social factors [9,21].

This manuscript deals with the learning space typology that considers social interactions as a means of learning. Regarding educational buildings, the literature has already identified that elements such as satisfaction or comfort, functionality, the possibilities of social interaction and place attachment are key for the development of learning [22]; this will be considered in this research as part of the social dimension of space.

1.1. Peer Effect and Active Methodologies in Higher Education

In recent decades, peer effect studies have provided contradictory results, including positive and negative influences [23,24] as well as large or small effects in similar contexts. Some investigations have focused on the characteristics of classmates. Booij, Leuven, and Oosterbeek [25] found that low-ability students perform better when in groups with peers of a similar skill level. Others have focused on group size, such as Brady, Insler, and Rahman [26], who identified different social influences depending on group size, showing negative effects at a broader company level and positive effects at a narrower company level.

The influences of social interactions and the peer effect have been analyzed in recent studies on educational buildings that include disruptive and attractive methodologies such as gamification [27,28]. In addition, classroom design can foster interaction and collaboration among peers, affecting teaching methodology and improving learning [29]. Specifically, flexible spaces are more appropriate for adapting to different teaching methodologies, including a better flow of interactions between users [30].

There is general agreement on the benefits of social relationships among classmates [31,32], since they provide companionship, affection, intimacy, assistance, improvement of self-esteem and emotional support, as a basis for the development of identity [33]. The results show that those students who participate in positive social interactions with other classmates are associated with greater academic motivation as well as a higher academic result [34,35].

1.2. Satisfaction and IEQ Perception

Comfort and satisfaction are social constructions that can influence the value of the indoor environment, not only over time, but also from one culture to another [36]. However, this satisfaction covers environmental aspects and social characteristics that can contribute to the mental harmony or instability of the users [37]. This indicator has been correlated with building characteristics, personal characteristics and variables related to the purpose of the space [38]. Under these premises, studies were carried out regarding buildings classified as "green", showing that the interior environments led to a positive perception that affected productivity [39,40]. In addition, some research has focused on how the social influence of friends and family affects the opinion and satisfaction of sustainable elements [41]. Non-physical and subjective aspects influence the way occupants perceive environmental comfort; therefore, psychological and social factors can positively affect users' perception of comfort [42]. For this reason, satisfaction as a variable that favors social relations is part of Kopec's theory of integration [43] on the relationship between human beings and space. The literature on buildings destined for education reiterates that the satisfaction of students with their environment is related to their academic results [44,45].

Studies on the comfort and satisfaction in school buildings have identified that a good quality environment positively affects the well-being, learning capacity and comfort of students [46]. It has been suggested that a better understanding of students' perceptions is necessary to understand their comfort level with the different variables of building design, such as temperature or lighting [47]. The effects of artificial light on the emotional state of adolescent students have also been explored, since inadequate lighting can be very harmful to the psyche of young people. Thus, ethical and healthy regulations regarding the optimization of lighting have been put forward [48]. Satisfaction regarding the indoor environment of schools, according to thermal comfort, air quality, and visual and acoustic comfort, has been addressed in recent studies, verifying that the discomfort of a specific element does not result in general discomfort; thus, individualized treatment of IEQs is necessary [49]. The literature has brought to light visual or aesthetic satisfaction, beyond the color of the classroom, as being influenced by images in primary education settings [50]. This is an unusual practice in university classrooms, but it is important to keep in mind the possible relationship with the place attachment. Other studies have shown that the level of satisfaction decreased when there were many people in the same room, which can be attributed to a lower degree of perceived control and greater necessary social interactions [51].

However, perceptions of comfort and satisfaction are usually incomplete or biased, which leads to failure when performing any type of intervention. Specifically, in educational buildings, the approach that involves students in POE provides researchers with contextualized information on which elements are most influential in overall comfort. This helps analyses to be carried out with greater precision, taking into account the factors that maximize solutions [13,52].

1.3. Place Attachment

In recent years, researchers have become more interested in the human dimension of sustainable design as it relates to health and well-being [53]. POE studies identified a series of outcomes related to the well-being of users, such as reduced absenteeism and stress, greater comfort and learning outcomes, and more positive attitudes [54,55]. However, among these human factors, place attachment and the relationships between people and their places have received little attention in the literature [56]. This fact is reflected in the multiple definitions of this construct in the literature before Scannell and Gifford [57] synthesized them and created an organizational framework with three main dimensions.

The first focuses on a personal and cultural dimension and is centered on who is becoming attached and how places are meaningful, both in individual and collective experience. The second brings a dimension that focuses on what a person is attached to, including physical and social characteristics, such as the natural environment that surrounds him or her or the opportunities for interaction with the rest of the users. The third, a psychological process dimension that focuses on how attachment includes certain behaviors, affective bonds and cognition, such as memories. Thus, in the case of students, their need to define their territory and their sense of belonging to it can be seen, for example, in the choice of seating area [58]. Affection is a key element in the process of creating a bond between the person and the place. Therefore, place attachment is more likely to occur in spaces with physical characteristics that support stress reduction, that evoke memories of people, and that facilitate the inclusion and interaction of other people [59,60]. It is also related to the personal or cultural circumstances of the users, which can lead to variations in the affective bonds with the different architectural contexts and even with the other users of the space [61].

Several studies have highlighted the value of place attachment concerning green or sustainable buildings and have tried to determine a connection between these feelings and pro-environmental behavior [62]. Thus, its consideration in educational buildings has the potential to support sustainable behavior, providing an incentive for green building practices [63]. Building design strategies focused on human attachment have also been found to improve community well-being, quality of life, and resilience [64]. They can also increase the amount of time spent in the building and the kinds of activity engaged in [65]. In this sense, Heerwagen and Zagreus [66] found an association between the feeling of place attachment and pride in the adoption of actions focused on sustainability. The results provided information on a series of psychosocial benefits, such as a more positive work experience, better communication between colleagues, and a strong connection with the environment and the company.

Regarding learning spaces, holistic studies on place attachment have shown that it has a greater value than other common IEQs, such as light, in the development of educational activities [67].

1.4. Objective

The objective of this research is to explore whether the social dimension of physical space in educational settings explains a student's academic results through a post-occupancy evaluation design.

2. Materials and Methods

The research was conducted using a quantitative approach that sought to understand the perception of students regarding their satisfaction with their indoor environment and its ability to foster learning interactions.

First, appointments were arranged with each dean or person in charge of infrastructure to visit centers to identify the learning space designs. Figure 1 shows selected classrooms and their diversity in layout design, lighting typology, furniture, information technology support, etc. Then, the professors who teach in these classrooms were identified, and they were contacted to determine their availability for applying the data collection instrument to their students. The purpose of the study and the average time of responding to the questionnaire were indicated. Consequently 21 of the 30 groups were able to establish a date to conduct the test.

The Learning Environment and Social Interaction scale (LESI) was provided in hard copy, and the students' answers were entered into a digital spreadsheet (educational Microsoft Excel 365).

The analysis of data consisted of a description of the sample, the reliability and validity, and prediction of the Grade Point Average (GPA) from IEQ satisfaction and learning interaction. First, the mean value and the standard deviation were defined to determine the empirical framework. Then, Cronbach's Alpha and Exploratory Factor Analysis were conducted to evidence the reliability and the construct validity of the data collection instrument. Finally, multiple linear regression was calculated to identify which items of the scale could predict academic performance and to what extent. The linear independence of predictor variables and the homoscedasticity of residuals were checked.

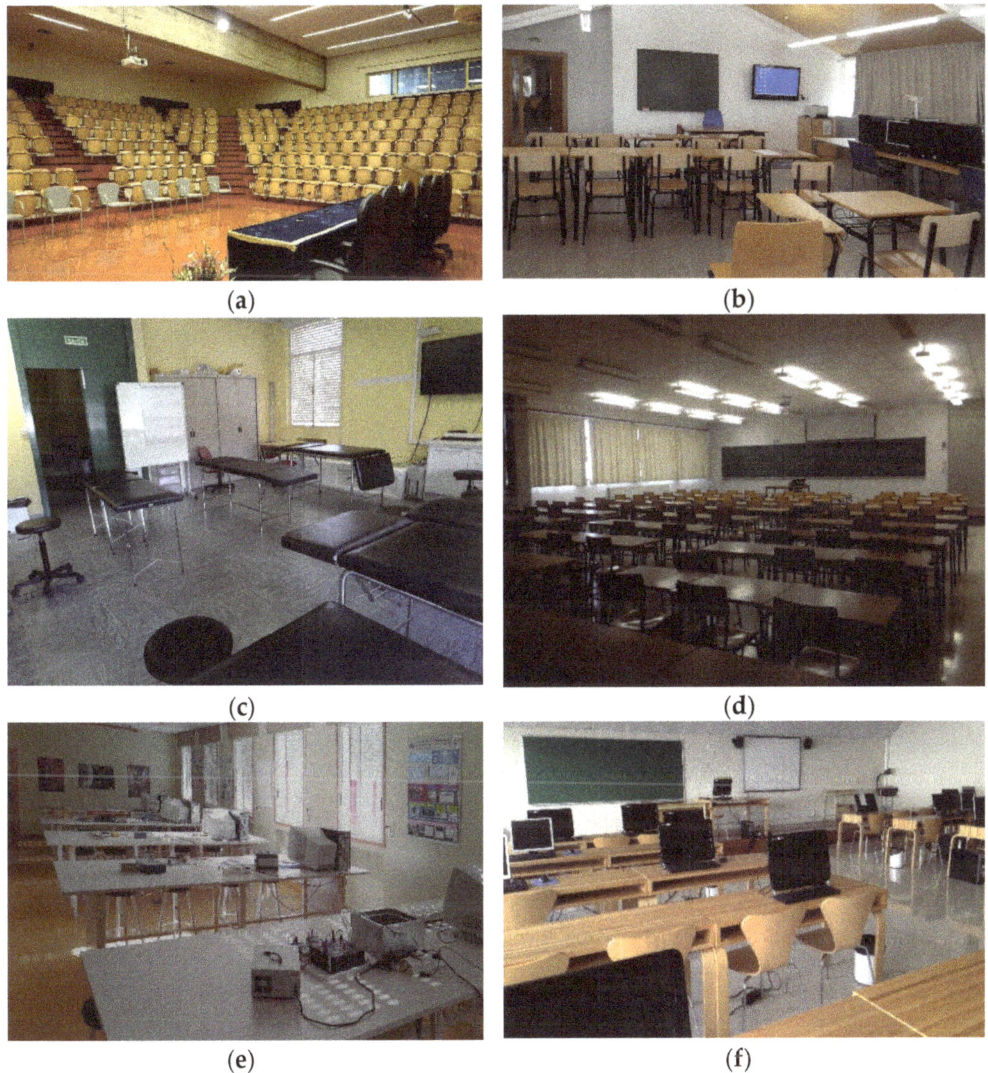

Figure 1. Pictures of some classrooms in the sample: (**a**) lecture hall; (**b**) practice room; (**c**) health practice room; (**d**) lecture room; (**e**) engineering practice room; (**f**) computer lab.

2.1. Sample

The sampling model is non-probabilistic by convenience and intentional, based on the representativeness of the areas of knowledge and the diversity of the space design. In other words, more and less illuminated rooms, more and less ventilated rooms, etc., were searched to obtain a sample with a wide range of possibilities in order to be able to relate the different levels of the predictor variables. The LESI scale was completed by 796 undergraduate students from 18 bachelor degree programs at Universidade da Coruña, who were reasonably balanced among arts and humanities, engineering and architecture, health science, science, and social and legal sciences. Table 1 shows the distribution of participants by bachelor degree program. The number of students in each degree program is proportional to participants in the study.

Table 1. Sample distribution by bachelor degree program.

Bachelor Degree	n	n (%)
Architecture	35	4.40
Biology	26	3.30
Chemistry	35	4.40
Civil engineering	61	7.70
Computer science	30	3.80
Early childhood education	80	10.10
Economics	62	7.80
Economy	16	2.00
Humanities	29	3.60
Industrial design and product development	8	1.00
Law	117	14.70
Mechanical engineering	12	1.50
Nursing	28	3.50
Occupational therapy	34	4.30
Podiatry	15	1.90
Primary education	81	10.20
Public works engineering	11	1.40
Social education	9	1.10
Sociology	22	2.80
Speech therapy	71	8.90
Technical architecture	14	1.80
Total	796	100.00

2.2. Data Collection Instrument

Learning Environment and Social Interaction (LESI) is a 1–7 Likert scale that is part of the Student Perception Questionnaire of Learning Space [68]. The instrument seeks to measure the collective perception of the environment by users regarding its power to favor learning interactions, the indoor environment quality satisfaction, and the importance of learning interaction in education. Students rated 18 independent variables from 1 (completely disagree) to 7 (completely agree). In addition, grade point average was requested at the beginning of the test template (Table 2).

Table 2. LESI scale items.

Item	Variable
The lecture classroom favors teacher–student interactions	V1
The lecture classroom favors interactions between students	V2
The practice classroom favors teacher–student interactions	V3
The practice classroom favors interactions between students	V4
Classroom design encourages participation	V5
Learning space attachment	V6
Lighting satisfaction degree	V7
Ventilation satisfaction degree	V8
Thermal level satisfaction degree	V9
Wall color satisfaction degree	V10
Acoustics satisfaction level	V11
Room layout satisfaction	V12
Furniture comfort satisfaction	V13
Connection with nature satisfaction	V14
Importance of professor–student interactions	V15
Importance of interactions with professors from other courses	V16
Importance of interactions between students	V17
Importance of interactions with students from other courses	V18

3. Results

3.1. Descriptive Analysis

Regarding the descriptive analyses (see Table 3), the LESI items could be grouped around four values:

- A value close to 3.40 was determined for the ventilation satisfaction degree (m = 3.32), the importance of interactions with professors from other courses (m = 3.43), the thermal level satisfaction (m = 3.45) and the furniture comfort satisfaction (m = 3.52).
- Values close to 4 represent the learning space attachment (m = 3.80), the connection with nature satisfaction (m = 3.84), the room layout satisfaction (m = 4.04), the acoustic satisfaction degree (m = 4.08), the importance of interactions with students from other courses (m = 4.14), the lecture classroom favors teacher–student interactions (m = 4.30) and the practice classroom favors teacher–student interactions (m = 4.31).
- A score of over 4.70 was determined for the wall color satisfaction (m = 4.62), the lighting satisfaction (m = 4.69), the lecture classroom favors interactions between students (m = 4.76), the practice classroom favors interactions between students (m = 4.82) and the classroom design encourages participation (m = 4.82).
- The best scored items received values close to 5.50, including the importance of interactions between students (m = 5.48) and the importance of professor–student interactions (m = 5.51).

Table 3. Descriptive results of LELI scale.

Variable	Min.	Max.	Mean	Std. Dev.
V1	1	7	4.30	1.716
V2	1	7	4.76	1.721
V3	1	7	4.31	1.793
V4	1	7	4.82	1.762
V5	1	7	4.82	1.844
V6	1	7	3.80	1.737
V7	1	7	4.69	2.576
V8	1	7	3.32	1.626
V9	1	7	3.45	1.713
V10	1	7	4.62	1.823
V11	1	7	4.08	1.665
V12	1	7	4.04	1.696
V13	1	7	3.52	1.715
V14	1	7	3.84	1.677
V15	1	7	5.51	1.488
V16	1	7	5.48	1.496
V17	1	7	4.14	1.790
V18	1	7	3.43	1.786

3.2. Reliability and Sample Adequacy

The sample for internal consistency analysis was 796. The Cronbach's Alpha index was used to check the level of reliability. Table 4 shows the results of Cronbach's α for the LESI scale, obtaining appropriate results (0.809).

Table 4. LESI internal consistency.

Cronbach's Alpha	Number of Items
0.809	18

3.3. Exploratory Factor Analysis

Exploratory Factor Analysis (EFA) was applied using the Principal Components method and Varimax rotation for the LESI scale. Previously, the Kaiser–Meyer–Olkin Sample Adequacy Measure (KMO = 0.767) was performed (see Table 5), which showed

Table 8. Standardized coefficients and collinearity statistics: LESI variables on GPA.

Variable	Beta	t	Sig.	Tolerance	VIF
(Constant)		23.356	<0.001		
V5	0.137	3.816	<0.001	0.931	1.075
V10	0.134	3.632	<0.001	0.880	1.137
V18	−0.118	−3.234	0.001	0.902	1.109
V11	0.134	3.512	<0.001	0.828	1.208
V8	0.131	3.566	<0.001	0.886	1.129
V15	0.107	2.843	0.005	0.855	1.169
V3	−0.072	−1.986	0.047	0.919	1.088

Another of the assumptions to be checked is linearity; Figure 3 displays the values that predict our estimation with respect to the values of the regression residuals. The result confirms the assumption of homoscedasticity, since the variance is practically homogeneous for the entire range of values. This figure also demonstrates compliance with the principle of linearity, since there is no non-linear pattern in the data cloud.

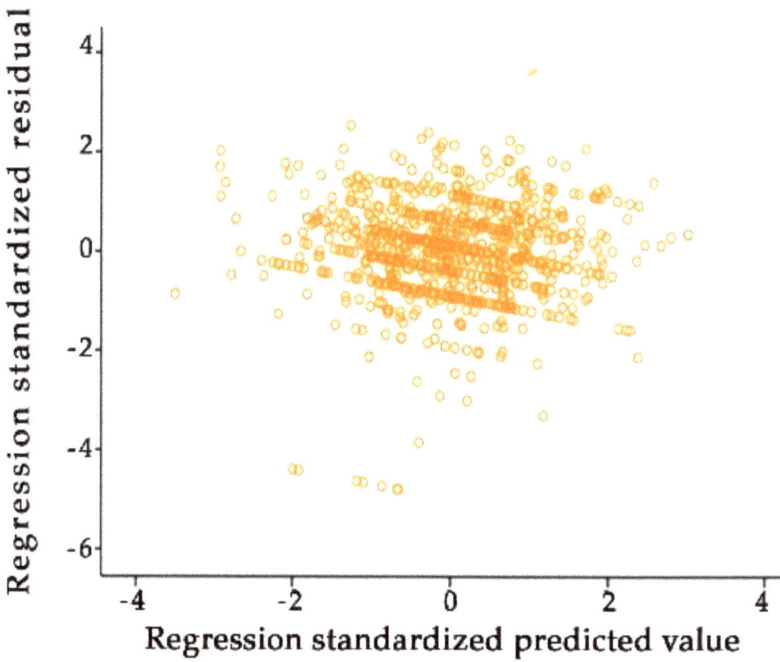

Figure 3. Cloud points of standardized predicted values vs. standardized residuals.

The last check requires that the distribution of the residuals follow a pattern close to normality. The P-P plot verifies compliance since, in general, the factors are close to or above the line (see Figure 4).

Once the validity of the model is verified, it is necessary to analyze the ANOVA results (see Table 9). This provides an F statistic value and an associated probability value, as well as sums of squares, degrees of freedom, and mean squares. A probability value less than 0.05 indicates that the model is consistent, thus allowing us to explain the relationship between the input and output variables.

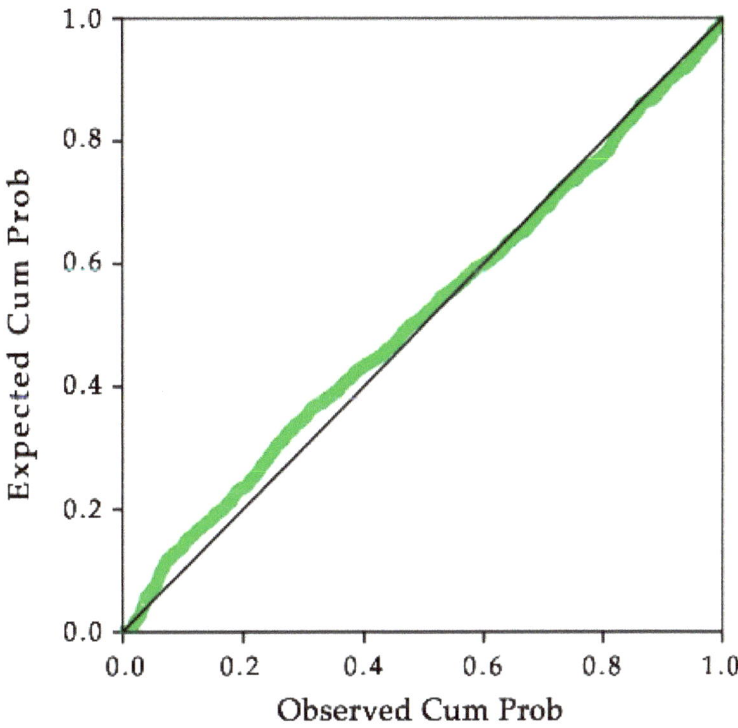

Figure 4. Normal P-P plot of Regression Standardized Residual. Dependent variable: GPA.

Table 9. ANOVA results.

Model	Sum of Squares	Difference	Mean Square	F	Sig.
Regression	67.056	7	9.579	9.990	<0.001
Residual	729.759	761	0.959		
Total	796.815	768			

Finally, Figure 5 shows the partial regression graphs of each variable in the model, where the line is the equation obtained from the linear regression analysis.

The multiple linear regression analysis performed on LESI demonstrated the existence of a relationship with academic outcome. The coefficient of determination was 0.076, while the mean square error was 0.9540.

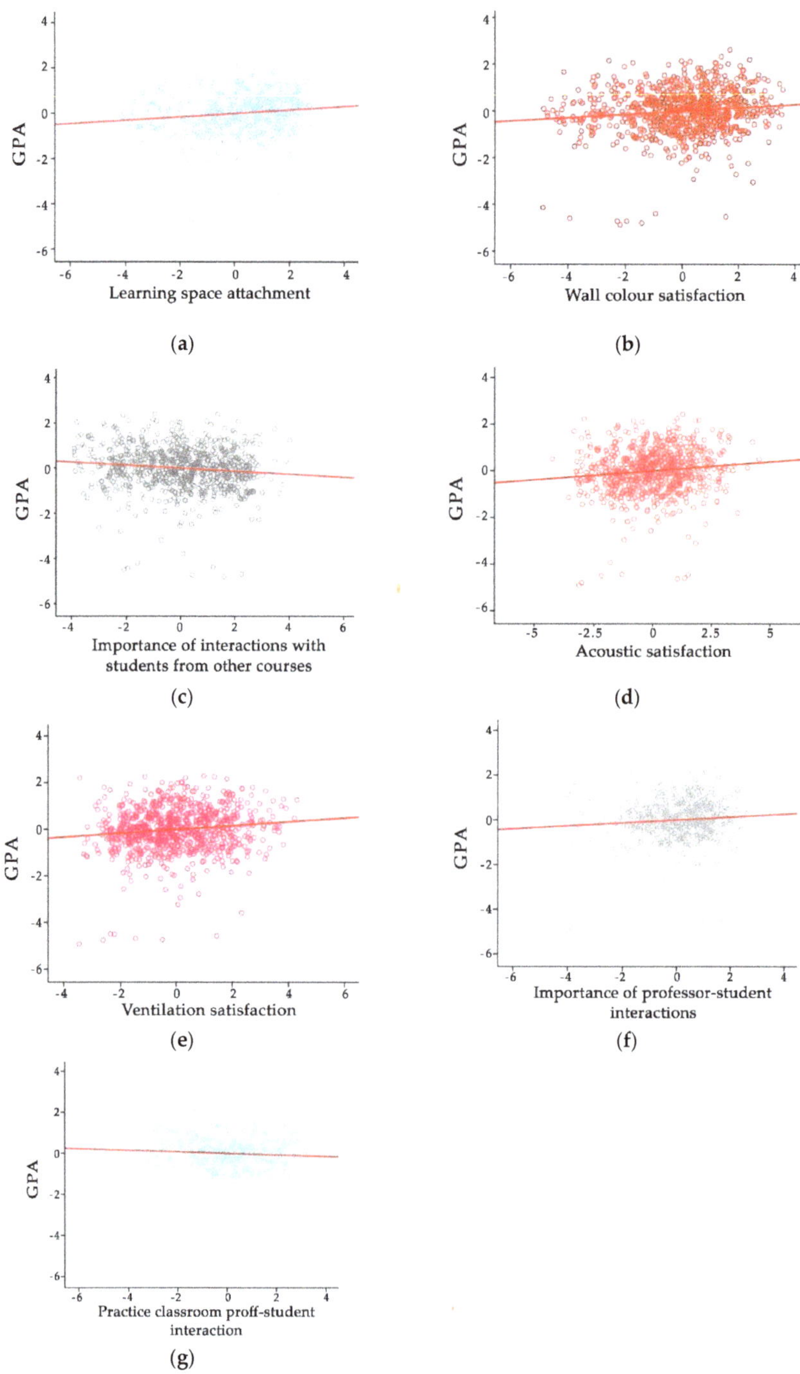

Figure 5. Partial regression graphs: LESI variables on GPA. (**a**) Learning space attachment; (**b**) Wall color satisfaction degree; (**c**) Importance of interactions with students from other courses; (**d**) Acoustics satisfaction level; (**e**) Ventilation satisfaction degree; (**f**) Importance of professor–student interactions; (**g**) Practice classroom favors teacher–student interactions.

4. Discussion and Conclusions

Sustainable building design is one of the priorities for the preservation of resources and energy. In recent decades, research based on post-occupation evaluation studies of indoor environment quality indicators have become more and more common. Factors such as geographic, cultural and climatological diversity have confirmed that it is not possible to develop a single model, but rather it is essential to disseminate research to diagnose reality. In addition, it is important to consider the functions, uses and habits in order to determine a sustainable design. In the case of educational buildings, in addition to IEQs, teaching methods should be taken into consideration. At present, several university models involve active teaching, which promotes the participation and interaction of students as a basis for learning. This research aimed to delve into three constructs—the peer effect, place attachment and IEQ satisfaction—and their relationship with academic outcomes.

For this, the Learning Environment and Social Interaction scale was designed and validated, which evidenced a structure of five factors: classroom design as a facilitator of social interactions, working design satisfaction, the learning value of social interaction, classroom environmental satisfaction and place attachment. Regarding the first factor, the literature shows that flexible designs allow teachers to promote greater learning and better adaptability to active methodologies, which also allows an improvement in the flow of social interactions between students [29,30]. The IEQ satisfaction has was divided into two factors: workspace design and classroom environmental satisfaction. The literature does not really show this separation, but there is agreement on the variables related to the activity performed in the space and those related to the environment [38,69]. Regarding the learning value of social interaction, positive relationships between classmates have shown benefits for academic performance and motivation [34,35]. Finally, place attachment has been confirmed as a factor in itself; as the basis of the link between the person and the place [60], it not only improves the well-being of the users of the building but also favors inclusion and interaction between people [57].

The regression results indicate that place attachment is the LESI variable that explains academic performance in the sample to the greatest extent. This contribution supports previous studies that demonstrated that place attachment had a higher value than other common IEQs, such as lighting, regarding the development of academic performance [67]. Regarding IEQ satisfaction, wall color, acoustics and ventilation also evidenced a direct relationship with the academic outcome, in line with evidence on satisfaction with the learning environment [44,45] and quality of environment [46].

Two variables of the learning value of social interaction showed an inverse and a direct relationship with the learning outcome: the importance of interactions with students from other courses and the importance of professor–student interactions. These findings indicate that greater interaction with the teacher may be related to a better understanding of the objectives or content of the subject and consequently of the academic outcome, while greater interaction with students from other courses leads to lower solvency of the course. Although it seems a consistent result, it is common for students to lean on peers from other courses, as they create bonds of friendship beyond academic assignments. However, this scenario would require a larger study to determine whether the support of outsiders is correlated with fewer interactions in the classroom itself. Previous research has verified the negative effects of a greater number of classmates and interactions and the positive effects of interactions in a smaller group [26], which could identified as the academic group within the classroom.

Finally, only one of the variables of classroom design as a facilitator of social interaction showed an inverse relationship with the academic outcome: the practice classroom favors teacher–student interactions. This result shows an apparent contradiction, in line with the literature [23,24]. Previously, it came to light that those students who assign higher value to teacher–student interactions obtain better GPAs, and those users who perceive that the classroom favors these interactions obtain worse results. The research does not allow for determination of the reason for this difference, but it seems logical that those who value interaction for their learning use it for academic purposes, also known as "learning interaction". Meanwhile, those who indicate that the design favors interaction would use it for social purposes.

This leads to the conclusion that the social dimension of the physical space contributes to the explanation of a student's academic result. In addition, the research contributes to the design and validation of LESI, the scale that those responsible for higher education centers can apply to diagnose their particular scenario and consequently manage the pertinent modifications.

This research focused on quantitative methods, due to its exploratory nature. A multi-method approach would be beneficial to complement the theory on learning space satisfaction and social interaction in higher education. Furthermore, more research is necessary both in higher education and at other educational levels, so that the high costs of building can be justified by substantial data on sustainable architecture in terms of purpose adequacy. Likewise, research on each factor, allowing an in-depth understanding of the particular complexity of each variable, is required.

Author Contributions: Conceptualization, V.L.-C.; methodology, V.L.-C.; software, V.L.-P.; validation, V.L.-C. and V.L.-P.; formal analysis, V.L.-C. and V.L.-P.; writing—original draft preparation, V.L.-C. and V.L.-P.; writing—review and editing, V.L.-C. and V.L.-P. Both authors have read and agreed to the published version of the manuscript.

Funding: This research received no external funding.

Institutional Review Board Statement: Not applicable.

Informed Consent Statement: Informed consent was obtained from all subjects involved in the study. The participation was optional and entailed the treatment, analysis and publication of findings.

Data Availability Statement: Not applicable.

Conflicts of Interest: The authors declare no conflict of interest.

References

1.   Wong, L.T.; Mui, K.W.; Hui, P.S. A multivariate-logistic model for acceptance of indoor environmental quality (IEQ) in offices. *Build. Environ.* **2008**, *43*, 6. [CrossRef]
2.   Heinzerling, D.; Schiavon, S.; Webster, T.; Arens, E. Indoor environmental quality assessment models: A literature review and a proposed weighting and classification scheme. *Build. Environ.* **2013**, *70*, 210–222. [CrossRef]
3.   Leccese, F.; Rocca, M.; Salvadori, G.; Belloni, E.; Buratti, C. Towards a holistic approach to indoor environmental quality assessment: Weighting schemes to combine effects of multiple environmental factors. *Energy Build.* **2021**, *245*, 111056. [CrossRef]
4.   Galasiu, A.D.; Veitch, J.A. Occupant preferences and satisfaction with the luminous environment and control systems in daylit offices: A literature review. *Energy Build.* **2006**, *38*, 728–742. [CrossRef]
5.   Ma, K.W.; Wong, H.M.; Mak, C.M. A systematic review of human perceptual dimensions of sound: Meta-analysis of semantic differential method applications to indoor and outdoor sounds. *Build. Environ.* **2018**, *133*, 123–150. [CrossRef]
6.   Pigliautile, I.; Casaccia, S.; Morresi, N.; Arnesano, M.; Pisello, A.L.; Revel, G.M. Assessing occupants' personal attributes in relation to human perception of environmental comfort: Measurement procedure and data analysis. *Build. Environ.* **2020**, *177*, 106901. [CrossRef]
7.   Földváry Ličina, V.; Cheung, T.; Zhang, H.; de Dear, R.; Parkinson, T.; Arens, E.; Chun, C.; Schiavon, S.; Luo, M.; Brager, G.; et al. Development of the ASHRAE Global Thermal Comfort Database II. *Build. Environ.* **2018**, *142*, 502–512. [CrossRef]
8.   Khan, M.; Thaheem, M.J.; Khan, M.; Maqsoom, A.; Zeeshan, M. Thermal comfort and ventilation conditions in healthcare facilities—Part 2: Improving indoor environment quality (ieq) through ventilation retrofitting. *Environ. Eng. Manag. J.* **2021**, *19*, 2059–2075.
9.   López-Chao, V.; Lorenzo, A.A.; Martin-Gutiérrez, J. Architectural indoor analysis: A holistic approach to understand the relation of higher education classrooms and academic performance. *Sustainability* **2019**, *11*, 6558. [CrossRef]
10. Zuhaib, S.; Manton, R.; Griffin, C.; Hajdukiewicz, M.; Keane, M.M.; Goggins, J. An Indoor Environmental Quality (IEQ) assessment of a partially-retrofitted university building. *Build. Environ.* **2018**, *139*, 69–85. [CrossRef]
11. Preiser, W.F.E. Post-occupancy evaluation: How to make buildings work better. *Facilities* **1995**, *13*, 19–28. [CrossRef]
12. Li, Y.; Yang, L.; He, B.; Zhao, D. Green building in China: Needs great promotion. *Sustain. Cities Soc.* **2014**, *11*, 6. [CrossRef]
13. Domínguez-Amarillo, S.; Fernández-Agüera, J.; González, M.M.; Cuerdo-Vilches, T. Overheating in Schools: Factors Determining Children's Perceptions of Overall Comfort Indoors. *Sustainability* **2020**, *12*, 5772. [CrossRef]

14. Mallory-Hill, S.; van der Voordt, T.; van Dortmont, A. Evaluation of innovative workplace design in the Netherlands. In *Assessing Building Performance*; Routledge: London, UK, 2004.
15. Wang, C.; Zhang, F.; Wang, J.; Doyle, J.K.; Hancock, P.A.; Mak, C.M.; Liu, S. How indoor environmental quality affects occupants' cognitive functions: A systematic review. *Build. Environ.* **2021**, *193*, 107647. [CrossRef]
16. D'Oca, S.; Pisello, A.L.; De Simone, M.; Barthelmes, V.M.; Hong, T.; Corgnati, S.P. Human-building interaction at work: Findings from an interdisciplinary cross-country survey in Italy. *Build. Environ.* **2018**, *132*, 147–159. [CrossRef]
17. Amasyali, K.; El-Gohary, N.M. Energy-related values and satisfaction levels of residential and office building occupants. *Build. Environ.* **2016**, *95*, 251–263. [CrossRef]
18. Ma, G.; Lin, J.; Li, N.; Zhou, J. Cross-cultural assessment of the effectiveness of eco-feedback in building energy conservation. *Energy Build.* **2017**, *134*, 329–338. [CrossRef]
19. Chen, C.; Hong, T.; de Rubens, G.Z.; Yilmaz, S.; Bandurski, K.; Bélafi, Z.D.; De Simone, M.; Bavaresco, M.V.; Wang, Y.; Liu, P.; et al. Culture, conformity, and carbon? A multi-country analysis of heating and cooling practices in office buildings. *Energy Res. Soc. Sci.* **2020**, *61*, 101344. [CrossRef]
20. Day, J.K.; McIlvennie, C.; Brackley, C.; Tarantini, M.; Piselli, C.; Hahn, J.; O'Brien, W.; Rajus, V.S.; De Simone, M.; Kjærgaard, M.B.; et al. A review of select human-building interfaces and their relationship to human behavior, energy use and occupant comfort. *Build. Environ.* **2020**, *178*, 106920. [CrossRef]
21. Schweiker, M.; Ampatzi, E.; Andargie, M.S.; Andersen, R.K.; Azar, E.; Barthelmes, V.M.; Berger, C.; Bourikas, L.; Carlucci, S.; Chinazzo, G.; et al. Review of multi-domain approaches to indoor environmental perception and behaviour. *Build. Environ.* **2020**, *176*, 106804. [CrossRef]
22. Cornell, P. The impact of changes in teaching and learning on furniture and the learning environment. *New Dir. Teach. Learn.* **2002**, *2002*, 33–42. [CrossRef]
23. Carrell, S.E.; Fullerton, R.L.; West, J.E. Does Your Cohort Matter? Measuring Peer Effects in College Achievement. *J. Labor Econ.* **2009**, *27*, 439–464. [CrossRef]
24. Carrell, S.E.; Sacerdote, B.I.; West, J.E. From Natural Variation to Optimal Policy? The Importance of Endogenous Peer Group Formation. *Econometrica* **2013**, *81*, 855–882. [CrossRef]
25. Booij, A.S.; Leuven, E.; Oosterbeek, H. Ability Peer Effects in University: Evidence from a Randomized Experiment. *Rev. Econ. Stud.* **2016**, *84*, 547–578. [CrossRef]
26. Brady, R.R.; Insler, M.A.; Rahman, A.S. Bad Company: Understanding negative peer effects in college achievement. *Eur. Econ. Rev.* **2017**, *98*, 144–168. [CrossRef]
27. López-Chao, V.; Lorenzo, A.A.; Saorín, J.L.; De La Torre-Cantero, J.; Melián-Díaz, D. Classroom indoor environment assessment through architectural analysis for the design of efficient schools. *Sustainability* **2020**, *12*, 2020. [CrossRef]
28. Ebersbach, M., Brandenburger, I. Reading a short story changes children's sustainable behavior in a resource dilemma. *J. Exp. Child Psychol.* **2020**, *191*, 104743. [CrossRef]
29. Attai, S.L.; Reyes, J.C.; Davis, J.L.; York, J.; Ranney, K.; Hyde, T.W. Investigating the impact of flexible furniture in the elementary classroom. *Learn. Environ. Res.* **2021**, *24*, 153–167. [CrossRef]
30. McArthur, J.A. Matching Instructors and Spaces of Learning: The impact of classroom space on behavioral, affective and cognitive learning. *J. Learn. Spaces* **2015**, *4*, 16.
31. Martin, A.J.; Dowson, M. Interpersonal Relationships, Motivation, Engagement, and Achievement: Yields for Theory, Current Issues, and Educational Practice. *Rev. Educ. Res.* **2009**, *79*, 327–365. [CrossRef]
32. Ryan, A.M. Peer Groups as a Context for the Socialization of Adolescents' Motivation, Engagement, and Achievement in School. *Educ. Psychol.* **2000**, *35*, 101–111. [CrossRef]
33. Furman, W.; Buhrmester, D. Age and Sex Differences in Perceptions of Networks of Personal Relationships. *Child Dev.* **1992**, *63*, 103. [CrossRef] [PubMed]
34. Furrer, C.; Skinner, E. Sense of relatedness as a factor in children's academic engagement and performance. *J. Educ. Psychol.* **2003**, *95*, 148–162. [CrossRef]
35. Liem, A.D.; Lau, S.; Nie, Y. The role of self-efficacy, task value, and achievement goals in predicting learning strategies, task disengagement, peer relationship, and achievement outcome. *Contemp. Educ. Psychol.* **2008**, *33*, 486–512. [CrossRef]
36. Humphreys, M.A. Quantifying occupant comfort: Are combined indices of the indoor environment practicable? *Build. Res. Inf.* **2005**, *33*, 317–325. [CrossRef]
37. Sant'Anna, D.O.; Dos Santos, P.H.; Vianna, N.S.; Romero, M.A. Indoor environmental quality perception and users' satisfaction of conventional and green buildings in Brazil. *Sustain. Cities Soc.* **2018**, *43*, 95–110. [CrossRef]
38. Schiavon, S.; Altomonte, S. Influence of factors unrelated to environmental quality on occupant satisfaction in LEED and non-LEED certified buildings. *Build. Environ.* **2014**, *77*, 148–159. [CrossRef]
39. Leaman, A.; Thomas, L.; Vandenberg, M. "Green" buildings: What Australian users are saying. *EcoLibrium* **2007**, *6*, 22–30.
40. Holmgren, M.; Kabanshi, A.; Sörqvist, P. Occupant perception of "green" buildings: Distinguishing physical and psychological factors. *Build. Environ.* **2017**, *114*, 140–147. [CrossRef]
41. Salazar, H.A.; Oerlemans, L.; van Stroe-Biezen, S. Social influence on sustainable consumption: Evidence from a behavioural experiment. *Int. J. Consum. Stud.* **2013**, *37*, 172–180. [CrossRef]

42. Castaldo, V.L.; Pigliautile, I.; Rosso, F.; Cotana, F.; De Giorgio, F.; Pisello, A.L. How subjective and non-physical parameters affect occupants' environmental comfort perception. *Energy Build.* **2018**, *178*, 107–129. [CrossRef]

43. Kopec, D.A. *Environmental Psychology for Design*; Fairchild: New York, NY, USA, 2006.

44. Temple, P.H. Learning spaces in higher education: An under-researched topic. *Lond. Rev. Educ.* **2008**, *6*, 229–241. [CrossRef]

45. Hopland, A.O.; Nyhus, O.H. Does student satisfaction with school facilities affect exam results? *Facilities* **2015**, *33*, 760–774. [CrossRef]

46. Teli, D.; Jentsch, M.F.; James, P.A.B. Naturally ventilated classrooms: An assessment of existing comfort models for predicting the thermal sensation and preference of primary school children. *Energy Build.* **2012**, *53*, 166–182. [CrossRef]

47. Teli, D.; James, P.A.B.; Jentsch, M.F. Investigating the principal adaptive comfort relationships for young children. *Build. Res. Inf.* **2015**, *43*, 371–382. [CrossRef]

48. Baeza Moyano, D.; San Juan Fernández, M.; González Lezcano, R.A. Towards a Sustainable Indoor Lighting Design: Effects of Artificial Light on the Emotional State of Adolescents in the Classroom. *Sustainability* **2020**, *12*, 4263. [CrossRef]

49. Korsavi, S.S.; Montazami, A.; Mumovic, D. The impact of indoor environment quality (IEQ) on school children's overall comfort in the UK; a regression approach. *Build. Environ.* **2020**, *185*, 107309. [CrossRef]

50. Alonso Sanz, A.; Jardón, P.; Lifante Gil, Y. The role of the classroom's images. Study of visual culture at three schools. *Vis. Stud.* **2019**, *34*, 107–118. [CrossRef]

51. Hedge, A.; Burge, P.S.; Robertson, A.S.; Wilson, S.; Harris-Bass, J. Work-related illness in offices: A proposed model of the "sick building syndrome". *Environ. Int.* **1989**, *15*, 143–158. [CrossRef]

52. Schweiker, M.; Wagner, A. The effect of occupancy on perceived control, neutral temperature, and behavioral patterns. *Energy Build.* **2016**, *117*, 246–259. [CrossRef]

53. Sharma, A.; Saxena, A.; Sethi, M.; Shree, V. Life cycle assessment of buildings: A review. *Renew. Sustain. Energy Rev.* **2011**, *15*, 871–875. [CrossRef]

54. Dreyer, B.C.; Coulombe, S.; Whitney, S.; Riemer, M.; Labbé, D. Beyond Exposure to Outdoor Nature: Exploration of the Benefits of a Green Building's Indoor Environment on Wellbeing. *Front. Psychol.* **2018**, *9*, 1583. [CrossRef]

55. Thatcher, A.; Milner, K. Is a green building really better for building occupants? A longitudinal evaluation. *Build. Environ.* **2016**, *108*, 194–206. [CrossRef]

56. Low, S.M.; Altman, I. Place Attachment. In *Place Attachment*; Springer US: Boston, MA, USA, 1992; p. 12.

57. Scannell, L.; Gifford, R. Defining place attachment: A tripartite organizing framework. *J. Environ. Psychol.* **2010**, *30*, 10. [CrossRef]

58. Kaya, N.; Burgess, B. Territoriality. *Environ. Behav.* **2007**, *39*, 859–876. [CrossRef]

59. Tester, G.; Ruel, E.; Anderson, A.; Reitzes, D.C.; Oakley, D. Sense of Place among Atlanta Public Housing Residents. *J. Urban Health* **2011**, *88*, 436–453. [CrossRef]

60. Scannell, L.; Gifford, R. The experienced psychological benefits of place attachment. *J. Environ. Psychol.* **2017**, *51*, 256–269. [CrossRef]

61. Tang, I.-C.; Sullivan, W.C.; Chang, C.-Y. Perceptual Evaluation of Natural Landscapes. *Environ. Behav.* **2015**, *47*, 595–617. [CrossRef]

62. Qu, Y.; Xu, F.; LYU, X. Motivational place attachment dimensions and the pro-environmental behaviour intention of mass tourists: A moderated mediation model. *Curr. Issues Tour.* **2019**, *22*, 197–217. [CrossRef]

63. Ramkissoon, H.; Weiler, B.; Smith, L.D.G. Place attachment and pro-environmental behaviour in national parks: The development of a conceptual framework. *J. Sustain. Tour.* **2012**, *20*, 257–276. [CrossRef]

64. Cole, L.B.; Coleman, S.; Scannell, L. Place attachment in green buildings: Making the connections. *J. Environ. Psychol.* **2021**, *74*, 101558. [CrossRef]

65. Hidalgo, M.C.; Hernández, B. Place attachment: Conceptual and empirical questions. *J. Environ. Psychol.* **2001**, *21*, 273–281. [CrossRef]

66. Heerwagen, J.; Zagreus, L. *The Human Factors of Sustainable Building Design: Post Occupancy Evaluation of the Philip Merrill Environmental Center*; U.S. Department of Energy, Center for the Built Environment, University of California: Berkeley, CA, USA, 2005.

67. Barrett, P.; Davies, F.; Zhang, Y.; Barrett, L. The Holistic Impact of Classroom Spaces on Learning in Specific Subjects. *Environ. Behav.* **2017**, *49*, 425–451. [CrossRef] [PubMed]

68. López-Chao, V. El Impacto del Diseño del Espacio Y Otras Variables Socio-Físicas en el Proceso de Enseñanza-Aprendizaje, Universidade da Coruña. Ph.D. Thesis, Universidade da Coruña, Coruña, Spain, January 2017.

69. Muñoz Cantero, J.M.; García Mira, R.; López-Chao, V. Influence of physical learning environment in student's behavior and social relations. *Anthropologist* **2016**, *25*, 249–253. [CrossRef]

sustainability

Article

Poor Ventilation Habits in Nursing Homes Have Favoured a High Number of COVID-19 Infections

Gastón Sanglier-Contreras, Eduardo J. López-Fernández and Roberto Alonso González-Lezcano *

Architecture and Design Department, Escuela Politécnica Superior, Campus Montepríncipe, Universidad San Pablo CEU, CEU Universities, 28668 Boadilla del Monte, Madrid, Spain; sanglier.eps@ceu.es (G.S.-C.); eduardojose.lopezfernandez@ceu.es (E.J.L.-F.)
* Correspondence: rgonzalezcano@ceu.es

Abstract: Residents of nursing homes have been significantly affected by COVID-19 in Spain. The factors that have contributed to the vulnerability of this population are very diverse. In this study, physical agents, chemical pollutants, population density and different capacities of residences were analysed to understand their influence on the number of elderly people who have died in geriatric centres in different autonomous communities (AACCs) of Spain. A statistical analysis was carried out on the variables observed. The results show that many residences with a larger number of deaths were private, with some exceptions. Physical agents and pollutants were found to be determining factors, especially for the communities of Extremadura and Castilla–La Mancha, although the large number of factors involved makes this study complicated. The compromise between air quality and energy efficiency is of great importance, especially when human health is at stake.

Keywords: quality air; epidemiology; data analysis; statistics; nursing homes; COVID-19

Citation: Sanglier-Contreras, G.; López-Fernández, E.J.; González-Lezcano, R.A. Poor Ventilation Habits in Nursing Homes Have Favoured a High Number of COVID-19 Infections. *Sustainability* **2021**, *13*, 11898. https://doi.org/10.3390/su132111898

Academic Editor: Giouli Mihalakakou

Received: 15 July 2021
Accepted: 22 October 2021
Published: 28 October 2021

Publisher's Note: MDPI stays neutral with regard to jurisdictional claims in published maps and institutional affiliations.

1. Introduction

Today, in developed countries, 80–90% of people's time is spent indoors [1], especially in their homes. The duration spent in the home varies between 60% and 90% of the day, and 30% of the time is spent sleeping [2,3]. Since homes contain air that is inhaled, the greatest exposure to potential air pollutants is in these interior spaces. The indoor environment of the home should facilitate rest and recovery [4,5]; therefore, as poor indoor air quality (IAQ) has harmful effects on health, it prevents these beneficial effects from being realised.

Since the energy crisis of the 1970s, buildings have become increasingly airtight, leading to the appearance of IAQ-related diseases, such as sick building syndrome (SBS) [6]. In addition, a relationship between air movement in buildings due to ventilation and the spread of infectious diseases has been demonstrated [7,8]. In this context, the benefits of indoor air exchange have been confirmed, although the influence of ventilation on the spread of infectious diseases is not clear [6]. Consequently, since the influence of airflow rates on health has not been quantified, the ventilation rates specified in different regulations are usually set according to comfort criteria (perceived conditions) [9–12].

However, a healthy indoor environment can be achieved by applying strategies necessary to improve the COVID-19 pollutant IAQ, which, in addition to increasing the supply of fresh air, include controlling pollution from emission sources, cleaning the air and improving the efficiency of ventilation [13–15]. Therefore, the indoor air quality (IAQ), especially in indoor residential spaces, has a strong influence on human health; thus, it is essential to design adequate ventilation, which ensures good IAQ since the main purpose of ventilation is to dilute or remove indoor contaminants by providing outdoor air [16–21].

The values set for air renewal to ensure comfort and eliminate odours have been modified over the course of history, according to variations in ventilation theories. Currently, the Basic Document HS 3 for indoor air quality, included in the Technical Building Code [22], provides data on minimum ventilation rates for residential buildings in Spain, depending

95

on the room of the dwelling. In European countries, in addition to the EN 15251 standard (which should be used if no national standard is available), there are also state regulations that, as in the Spanish standard, provide data on minimum ventilation rates based mainly on body odours (with CO_2 as an indicator) and, to some extent, according to primary emissions from some building materials [23,24].

On the other hand, even if the regulations are complied with, indoor air quality may still be inadequate if stagnant air zones are generated, and therefore, health and comfort problems may develop [25–29]. In-depth research in this field is still needed; therefore, the debate on how much ventilation is sufficient to achieve good indoor air quality, capable of preventing both odours and the emergence and spread of diseases, is still ongoing.

Spain is one of the countries most affected by the COVID-19 pandemic. According to data provided by the Spanish Ministry of Health's Centre for the Coordination of Health Alerts and Emergencies [30], the number of deaths as of 16 May 2020 was 27,563, and 230,698 people were infected.

Based on the number of elderly people who have died in nursing homes [31], we find that the most affected autonomous communities (AACCs) were Castilla–La Mancha (14.44%), Extremadura (14.13%), Castilla y León (13.71%) and Aragón (12.90%), compared to the total number of deaths in these communities (elderly and non-elderly). These numbers are very similar to the number of the elderly who died in these communities: Castilla–La Mancha (17.38%), Extremadura (16.80%) and Aragón (15.36%).

According to data provided by the AACCs, the number of elderly people who have died as a result of COVID-19 in the approximately 5417 Spanish homes for the elderly, including public, subsidised and private residences, stands at 18,354 deaths. Only people who died after testing positive for coronavirus have been registered as casualties of COVID-19, i.e., tests have not been performed post-mortem, so those who were not tested are not listed as having died of coronavirus.

It is observed (Figure 1) that the number of private residences in AACCs is greater than the number of public residences in a proportion of 3 to 1, with the exception of the autonomous communities of Extremadura, Castilla–La Mancha and, to a lesser extent, Canarias.

The population pyramid in Spain continues to reflect population ageing, measured by the increase in the proportion of elderly persons, i.e., those aged 65 years old and over. According to the latest statistical data from the Continuous Register of the National Statistics Institute (INE), on 1 January 2019, there were 9,057,193 elderly people, 19.3% of the total population of 47,026,208 [32]. This population of elderly people continues to increase, both in number and proportion. The average age of the population, which is another way to measure this process, is 43.3 years; in 1970, it was 32.7.

Below is a comparative chart of the population over 65 years of age and the number of deaths in that age range for each AACC.

The data show that the communities of Madrid, Extremadura, Castilla–La Mancha and Baleares have been most affected by deceased elderly persons relative to the population of persons over 65 years of age in these communities (see Figures 1 and 2). On the other hand, Ceuta, Melilla, Canarias and Galicia have been less affected.

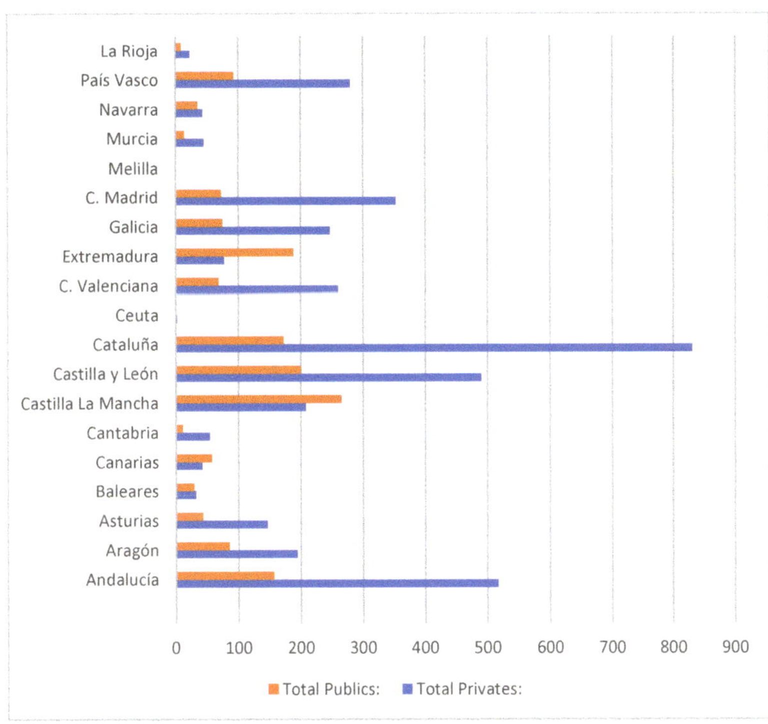

Figure 1. Distribution of the number of public and private nursing homes in the different autonomous communities of Spain.

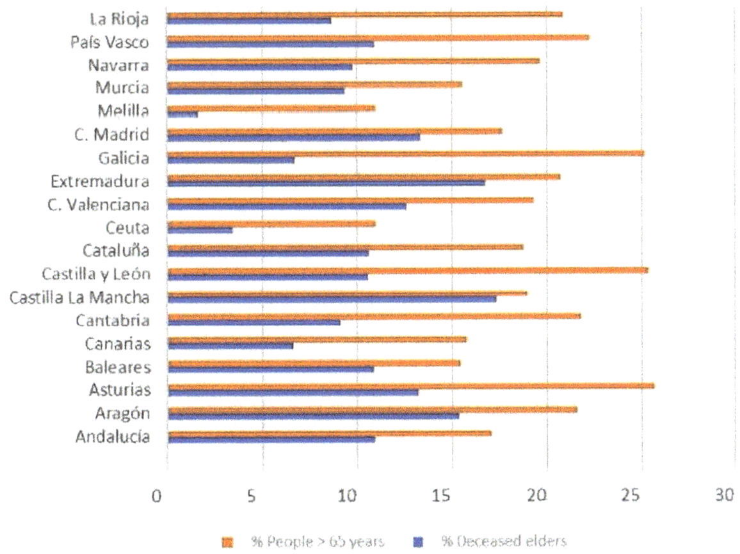

Figure 2. Graph of number of people >65 years old and deaths of people in this range by autonomous community.

The first study carried out in Europe on indoor air quality [33] includes detailed statistical data from the Eurostat Union Statistics on Income and Living Conditions. The data demonstrate a correlation between people's health and the conditions of the buildings in which they live.

In March of this year, the *European Respiratory Journal* published a study that revealed the importance of maintaining adequate indoor air quality conditions in residential buildings. Based on this study, it was deduced that older people are more likely to suffer negative effects on lung health due to indoor air pollution in buildings, compared to younger adults.

It has been shown that city dwellers (especially the elderly and children) spend between 58% and 78% of their time in an indoor environment that is polluted to a greater or lesser degree. This is a problem that has been exacerbated by the construction of buildings that are designed to be increasingly airtight and that recycle air with a smaller proportion of fresh air from outside in order to increase energy efficiency. Pollution of the indoor environments of buildings is the cause of many health problems of various kinds, which can range from simple fatigue or discomfort to symptoms consistent with allergies, infections and cancer, among others.

The pollutants present in the indoor air of buildings (residences), whether chemical, physical or biological, vary depending on the activities that take place in such spaces, the health of the occupants, the physical infrastructure of the building and its material assets and the quality of the surrounding air. At present, environmental pollutants, such as environmental tobacco smoke, formaldehyde, radon, mineral fibres, isocyanates and epoxy resins have been identified as some of the main emerging factors that can increase the risk of diseases, such as allergies, asthma, fertility disorders and cancer [34]. Physical factors that influence comfort are mainly related to relative humidity, average air speed, temperature and noise. In addition, there are chemical pollutants, including carbon dioxide (indicative of insufficient indoor air renewal), carbon monoxide, sulphur dioxide, volatile organic compounds, suspended particles, ozone, radon, etc., as well as various pathogens. In short, the population is faced daily with an array of pollutants not only in buildings, but also in the outside air, water, food, consumer products, etc. Determining the pollutant composition, daily exposure dose and interaction with the human body and the environment is quite a complex undertaking [34,35].

Volatile organic compounds (VOCs) play an important role in the process of assessing the IAQ. They negatively affect both the comfort and health of occupants [35,36]. The effects of VOCs in indoor spaces vary from sensory irritation at medium exposure levels to toxic effects at high exposure levels [37]. Formaldehyde (HCHO) is especially important, as it is known to be the most common irritant in indoor air, causing eye, nose and throat irritation (at concentrations above 0.1 mg/m^3) and may even cause asthma attacks at high concentrations [38,39].

Previous studies [40,41] provide total VOC concentration data, using the concept of total volatile organic compounds (TVOCs). This parameter is used to allow a simpler and faster assessment than the analysis of a high amount of VOCs, which are usually detected in indoor spaces [42]. The concentration of any pollutant in an indoor space is the result of a balance between the network in that space and what is removed or added by ventilation. Therefore, if the TVOC level is high, it indicates that there are significant sources of contamination or that local ventilation is inadequate. Therefore, it is important to measure TVOCs, as they are an indication of the pollution load in the indoor environment and the sustainability of the ventilation rate [43]. It should be noted that TVOCs do not include all VOCs present in indoor air, as some organic pollutants are excluded, such as low molecular weight aldehydes, e.g., HCHO, because the method of identification differs from that of other VOCs. Therefore, in order to characterise the IAQ and determine the adequate ventilation rate, it is essential to consider the concentration of VOCs and HCHs.

Therefore, it is difficult to assess the health risks (measurement, tolerance level, exposure time, effects, etc.) in the indoor environment; preventive and corrective maintenance in the facilities involved is important in order to promote healthy environments.

In this study, chemical pollutants (such as formaldehyde, nitrogen dioxide, ozone, PM_{10} and $PM_{2.5}$ particulate matter and sulphur dioxide), physical agents (such as temperature, humidity, precipitation and hours of sunshine) and socio-demographic variables (such as population density and the capacity of different public or private homes for the elderly) were analysed to understand their association with the indoor air quality of the homes and to determine whether they influence the health of the elderly who live there.

The analysis was carried out in two stages, first taking into account the sizes of different residences and subsequently evaluating the effects of physical agents and pollutants.

For the first stage, the statistical technique of multivariate correlation between the variables was applied in order to determine the relationships between the number of deaths in the residences and the 12 variables based on the size of the residence. Then, a cluster or conglomerate analysis was carried out to determine the relationship between the nursing homes and the autonomous communities under study.

2. Materials and Methods

In addition to children and pregnant women, other population groups are vulnerable to the effects of air pollution. These include people with respiratory diseases, such as asthma, chronic bronchitis or emphysema, those with cardiovascular diseases or diabetes, and elderly people with chronic diseases, particularly those living in residential homes.

The quality of the air in residences is in the IDA2 category, according to the Regulation of Thermal Installations (RITE) in Spain [24]. In this type of building, it is necessary to have a good quality of air in the indoor environment because the elderly population is more vulnerable to infectious agents.

Age and diseases (cardiovascular, respiratory problems) are factors that make the elderly more likely to be negatively affected by pollutants present in the air of residential environments. Susceptibility to pollutant effects is also influenced by the individual's lifestyle, diet, genetic predisposition, etc.

2.1. Comfort

In indoor environments, the ability to regulate temperature is provided by heating, ventilation and air-conditioning systems. The human body has the ability to regulate its temperature within a range of degrees. Thermal comfort in an indoor environment refers to the general sense of temperature and humidity felt by most people who live there.

Thermal comfort means that a person feels good from the perspective of the surrounding hygrothermal environment. Extreme thermal values can be harmful, even deadly, to human beings. This is because human beings are homeothermic, i.e., they must maintain certain vital parts at an approximately constant temperature. To achieve a feeling of thermal comfort, the most advisable condition is an ambient temperature that is slightly higher than the air temperature and a flow of radiant thermal energy that is the same in all directions and is not excessive above the head. In the UNE EN ISO 7730 standard (thermal comfort in moderate environments) [44], thermal comfort is given by the thermal balance between the body and the environment, i.e., a person will feel comfortable when the internal heat generated and the losses due to evaporation from the body are compensated by the losses or gains due to latent, sensitive or radiant heat with respect to the environment.

The following table (Table 1) shows data on the average temperature and humidity, precipitation, hours of sunshine [45] and the number of elderly people ($\geq$65 years) who have died in homes and in total for the different autonomous communities [30].

Table 1. Data on deceased elderly people and physical agents in the autonomous communities.

CC.AA.	(RD) Residential Deaths	(CD) Community Deaths	Average Temperature (C)	Average Humidity (%)	Precipitation (mm)	Sum Hours (h)
Andalucía	527	1355	19.4	53.6	344.5	3342
Aragón	704	838	14.6	61	272.7	3112
Asturias	192	313	13.8	80	908.2	1962
Baleares	84	216	19.3	73.5	466.5	3034
Canarias	18	151	22	61	78.5	3165
Cantabria	135	206	13.8	74	1157.6	1826
Castilla–La Mancha	2395	2883	16.4	49	221.8	2779
Castilla y León	2519	1940	12.1	65	384.3	2894
Cataluña	3394	5915	16.9	63.3	329.3	2731
Ceuta	0	4	19.1	74	650.2	2640
C. Valenciana	518	1365	19.1	62	335.6	2808
Extremadura	418	497	17.4	53	271.8	3326
Galicia	269	604	14.5	74	798.5	2342
C. Madrid	5909	8826	16	53	256.1	2970
Melilla	0	2	19.6	70	239.7	2778
Murcia	67	144	20	55	177.9	3348
Navarra	422	501	13.2	67	696.9	2512
País Vasco	584	1455	14.4	75.5	1456.7	1780
La Rioja	199	348	144	61	3907	2708

2.2. Air Quality Factors

The quality of the environment in nursing homes is affected by indoor air pollutants, such as dust, suspended particles, CO_2, CO, NOx, VOCs, bacteria, fungi and viruses, as well as pollution from outside. However, nursing homes need special maintenance of the facilities and environment (air renewal) for the following reasons:

- Elderly residents spend practically all of their time in these places (90%) since it is their home.
- The quality of the environment in the residences will be vitiated by a greater number of viruses, due to the fact that their occupants usually suffer from different infections.
- In residences, food and different types of drugs or medicines are maintained on the premises to care for the elderly, and these products need the air quality to be optimal.
- It is essential to use air-conditioning systems in homes to ensure the comfort of the users and to renew the indoor air. In winter, it is necessary to temper the indoor air to prevent cold currents, which can affect the health of the elderly. In summer, the opposite will occur, and air-conditioning systems need to be used to prevent the occurrence of hot flushes or heat in the occupants.
- The use of air-conditioning systems in residences is as important as their cleaning and maintenance: the lubrication of mechanisms, revision, change of filters, etc.
- Cleaning and disinfection of the building need to be carried out daily to maintain the quality of the indoor environment in the residence halls since different groups of people live together in these areas every day. In the residents' rooms, in the reception areas and in the corridors, it is necessary to carry out a deep cleaning or even disinfection several times a day to improve the safety of the guests and visitors.

2.3. Consequences of Poor Air Quality

The World Health Organization (WHO) warned that air pollution (outside air) kills about seven million people every year. Indoor pollution levels can be as much as 10–100 times higher than outdoor concentrations, and people (with the elderly being more vulnerable) exposed to poor-quality residential environments can suffer many health problems as a result:

- Airways: dryness, itching/heartburn, nasal congestion, sneezing and sore throat;
- Lungs: chest tightness, choking sensation, dry cough and bronchitis;
- Skin: redness, dryness and generalised itching;
- General malaise: headache, weakness, drowsiness/lethargy, difficulty concentrating, irritability, anxiety, nausea and dizziness;

- Diseases: hypersensitivity pneumonitis, humidification fever, asthma, rhinitis and dermatitis;
- Infections: legionellosis, Pontiac fever, tuberculosis, common cold and flu.

2.4. Chemical Pollutants

The pollutants studied in this work, as well as their impact on human health, are described below.

2.4.1. Nitrogen Oxides (NOx)

The presence of NOx is related to the burning of fuels, mobile sources (vehicles), industrial processes and some natural processes (lightning, and soil microorganisms). Combustion processes emit a mixture of nitric oxide (90%) and nitrogen dioxide (10%). In turn, nitric oxide reacts with other chemicals in the air to become nitrogen dioxide. In indoor environments, the main sources of NO_2 emissions are heating systems and gas stoves, as well as tobacco smoke.

At low concentrations, nitrogen oxides are irritating to the upper respiratory tract and eyes. Prolonged exposure can cause pulmonary oedema. Excessive exposure to nitrogen oxides can cause health effects on the blood, liver, lung and spleen. Nitrogen dioxide is also one of the gases that contribute to acid rain that damages vegetation and buildings and contributes to the acidification of lakes and streams.

2.4.2. Suspended Particles

These particles are usually referred to as total suspended particles (TSP) and include all particles with diameters ranging from less than 0.1 microns to 50 microns, as larger particles are deposited by gravity. TSP is expressed as PM (particulate matter) with a sub-index referring to particle diameter, and the unit is the weight of particles per volume of air (mg/m^3 or $\mu g/m^3$). The larger the particle size, the shorter the time they remain suspended in the air and the shorter the distances they can travel. Particles larger than 10 microns fall rapidly near the source that produces them; PM_{10} particles (with a diameter of ≤ 10 microns) can remain suspended for hours and travel from 100 m to 40 km, while $PM_{2.5}$ particles (with a diameter of ≤ 2.5 microns) can remain in the air for weeks and are capable of moving hundreds of kilometres, moving with air currents and penetrating premises through ventilation systems.

The main sources of particulate matter outdoors are road traffic, especially diesel vehicles, industrial processes, incinerators, quarries, mining, stack emissions, coal heating, etc. Other important sources of particulate matter are dust from agricultural work, road construction or vehicle traffic on unpaved roads. On the other hand, particulate matter is present in almost all indoor environments, mainly from combustion appliances and tobacco smoke. It can also have a biological origin, such as pollen, spores, bacteria and fungi. Typically, most particles of anthropogenic origin are in the range of 0.1–10 μ.

The size range that can be considered dangerous in relation to its effects on human health and air quality is between 0.1 microns and 10 microns in diameter since these particles, once inhaled, generally have a greater capacity to penetrate the respiratory system. PM_{10} particles are deposited in the upper respiratory tract (nose) and in the trachea and bronchi, while $PM_{2.5}$ particles with a smaller diameter can reach the bronchioles and alveoli of the lungs.

2.4.3. Formaldehyde

Formaldehyde is a major indoor air pollutant, and due to its chemical properties and serious health effects, an individual assessment is recommended. It is often present in the structure of modern building installations and furnishings, and its concentrations are higher indoors than outdoors. Urea–formaldehyde foam insulation (UFFI) was widely used in the construction of houses until the early 1980s, although its installation is now rare. The main sources of exposure to formaldehyde include particle board, varnishes, lacquers, glues, fibreglass, carpets, non-iron fabrics, paper products and certain cleaning and disinfection products. Due to the extremely high concentrations of formaldehyde in tobacco smoke, smoking is a major source of this compound. Gas stoves and ovens and open fireplaces are also sources of formaldehyde exposure.

Studies conducted in Canada since the early 1990s indicate the presence of formaldehyde in households in concentrations ranging from 2.5 $\mu g/m^3$ to 88 $\mu g/m^3$, with an average of between 30 $\mu g/m^3$ and 40 $\mu g/m^3$ [46].

The main form of exposure is inhalation; it can also be absorbed through skin contact. The main effects of acute exposure to formaldehyde are irritation of the conjunctiva of the eye and the mucosa of the upper and lower respiratory tract. The symptoms are temporary and depend on the level and duration of exposure. Exposure to high concentrations of formaldehyde may cause burns to the eyes, nose and throat. In the long term, exposure to moderate concentrations of formaldehyde (chronic exposure) may be associated with respiratory symptoms and allergic sensitivity, especially in children. Prolonged or repeated skin contact leads to irritation and dermatitis.

2.4.4. Ozone

In the indoor environment, the ozone originates from equipment that generates a potential discharge between metal plates or with the existence of ultraviolet radiation. This occurs in photocopiers, laser printers, electrostatic equipment for air purification, electric motors and equipment with UV radiation, such as those used in disinfection.

Due to its oxidising power, the immediate health effects are irritation of the respiratory tract and eyes, coughing, breathing difficulties, etc. In the medium term, there may be a general decrease in physical performance, as well as symptoms of general malaise, such as headache, tiredness, heaviness, etc. In the long term, it can produce alterations in pulmonary function (pneumonitis and pneumonia). In general, the effects of exposure to the ozone are accentuated by a higher concentration, longer duration of exposure and higher levels of activity during exposure, although the form of the dose—response relationship is not known. The severity of the response is strongly dependent on the sensitivity of the respiratory system and often on the health status of the exposed person.

2.4.5. Sulphur Dioxide

Data were collected on various pollutants in the different AACCs in Spain. The limit and objective reference values that appear in this study are those established by Directive 2008/50/EC [47] and Royal Decree 102/2011 [48], as well as those recommended by the World Health Organization (WHO) [49]. The data in the following table (Table 2) were extracted from the 127 zones and agglomerations established for the measurement of nitrogen dioxide in the Spanish territory, organised by AACC, with their respective measurement stations. The exceedances of the legal limits and WHO references by zone or agglomeration are reflected in the table. The values that appear correspond to the average value of all of the data collected by the stations in the zone (whether they exceed the limits or not). Some stations are the only representative of their area, and therefore, their data correspond to the average value of the area. The target value for the protection of human health from tropospheric ozone is set for a three-year period, in this case, for the years 2017, 2018 and 2019. The remaining pollutants refer to the year 2019 [50].

Table 2. Table of chemical contaminants with their limits according to regulations.

	PM_{10} (Particles < 10 Micras)		$PM_{2.5}$ (Particles < 2.5 Micras)		NO_2	O_3 (Tropospheric Ozone)			SO_2	Formal-dehyde
	Daily Value	Annual Average	Daily Value (OMS)	Annual Average	Annual Average	Eighth Hour (Normative)	Eighth Hour (OMS)	AOT40 (Normative)	Daily Value (OMS)	Annual Average
	No Days > 50 μg/m^3 Normative: Máx = 35 OMS: Máx:3	μg/m^3 Normative: Máx = 40 OMS: Máx:20	No Days > 50 μg/m^3 OMS: Máx:3	μg/m^3 Normative: Máx = 25 OMS: Máx:10	μg/m^3 Normative: Máx = 40 OMS: Máx:40	No Days > 120 μg/m^3 Normative: Máx = 25	No Days > 100 μg/m^3 OMS: Máx = 3	Normative: Máx: 18,000	No Days > 20 μg/m^3 OMS: Máx = 3	μg*m^2
Andalucía	5.36	23.27	12.36	13.27	15,9	18.36	94.18	18,767.54	0.36	59.2
Aragón	1	14.25	2.75	8.75	8	12.75	69.5	13,766.5	0	60.4
Asturias	7.75	22	6.25	11.75	13.5	1.75	18.75	3223.25	8.25	54.6
Baleares	1	13.75	1.33	7.33	11.75	8.75	77	14,018.5	0.75	76.1
Canarias	17	22.83	7.5	8.16	11.83	0.83	16.16	2274.66	2.66	57.4
Cantabria	4.33	19	1.66	9.3	18	1	35.33	4180.66	1.33	68
Castilla La Mancha	15.33	21.66	13.66	10.66	11.66	23.66	88	18,153.66	2.33	59.3
Castilla y León	3.1	15.2	1.8	6.4	10.1	9.6	66.1	10,345.5	2	57
Cataluña	2.2	19.6	5.64	10.5	15.92	14.66	77	19,245.78	1.76	58.2
Ceuta	N/A	N/A	N/A	N/A	N/A	N/A	N/A	N/A	N/A	61.3
C.Valenciana	1.21	14.78	4.28	9	11.71	16.64	90.28	19,060.35	0.64	61
Extremadura	5.5	15	4.5	9	11.71	16.64	90.28	19,060.35	0.64	61
Galicia	5	22	10.66	12.33	20.83	6.5	33.5	4911.16	2.66	65
C.Madrid	4.71	16.28	4	10	20.42	36.14	110.71	22,506.42	0.42	57.1
Melilla	N/A	N/A	N/A	N/A	N/A	N/A	N/A	N/A	N/A	60.6
Murcia	26.5	27	N/A	N/A	27	28.5	76.5	22,224.5	20.5	57.8
Navarra	0.5	13.5	5	12	11.25	5	29.75	8902	1.5	59.8
País Vasco	0.87	15	1.62	8.25	15.25	5	36	7419.25	0.28	58.6
La Rioja	2	17	1	9	7	7	59	10,237	0	60.1

For the interpretation of the data in Table 2, the limit values of the analysed pollutants are described below:

1. PM_{10} particles:

    - Daily value: No. of days in the year when the 50 μg/m^3 limit is exceeded. When it is greater than 35 days, the daily limit established by the regulations is exceeded, and if it is greater than 3 days, the WHO recommendation is exceeded.
    - Annual average: Average value of PM_{10} during the year. The limit established by the regulation is 40 μg/m^3 per year, while the WHO recommends not to exceed an annual average of 20 μg/m^3.

2. $PM_{2.5}$ particles:

    - Daily value: Number of days during the year when 25 μg/m^3 is exceeded. When it is greater than 3 days, the WHO recommendation is exceeded.
    - Annual average: Average value of $PM_{2.5}$ during the year. The regulations do not allow exceeding 25 μg/m^3 per year. WHO recommends not to exceed 10 μg/m^3 as an annual average.

3. Nitrogen dioxide NO_2:

    - Annual average: Average value of NO_2 during the year. The annual limit value set by the regulations is 40 μg/m^3, which is in line with the WHO recommendation.

4. Ozone O_3:

    - Eight-hour value: Number of days over the year when the average value of 120 μg/m^3 (legal) or 100 μg/m^3 (WHO) ozone is exceeded in 8 h periods (defined as the maximum daily 8 h moving average). The regulations do not allow more than 25 days per year (averaged over three consecutive years), a threshold also adopted in this report for the WHO recommendation (in 2018).
    - AOT40 May–July: Sum of the difference between hourly concentrations above 80 μg/m^3 and 80 μg/m^3 between 8:00 and 20:00, from 1 May to 31 July. The

legal target is 18,000 $\mu g/m^3$ h (the average of over five consecutive years), and the long-term target is 6000 $\mu g/m^3$ h (in 2019).

5. Sulphur dioxide SO_2:

- Daily value: Number of days per year when the average daily value of 125 $\mu g/m^3$ (legal) or 20 $\mu g/m^3$ (WHO) of SO2 is exceeded. The regulations do not allow more than 3 days per year, a threshold that is also adopted in this report for the WHO recommendation.

The Table 2 shows the average chemical contaminant measurements for each AACC and the limits according to the regulations specified above.

There are various types of buildings that require special air-conditioning and ventilation conditions, due to their unique use and the special sensitivity of their occupants to temperature changes or indoor air pollution, among other factors.

This is the case for homes for the elderly or geriatric centres, which, although not subject to the regulatory requirements for the specific temperature, relative humidity and ventilation conditions in hospitals, are buildings that require appropriate air-conditioning and ventilation systems; this is necessary to enable the elderly people who live in these facilities to achieve a good standard of living in terms of their personal well-being, due to their special needs in terms of health and comfort.

Air conditioning and other factors related to architecture and interior design can have a profound impact on the sense of home experienced by the elderly who live in this type of centre; furthermore, the dimensions of the room and the position of various elements within a room, such as furniture, have an influence on the movement of air and, therefore, on the efficiency of ventilation [51].

Considering the ventilation of a dwelling in general, research [52] has shown that for an uninhabited dwelling, the concentration of $PM_{2.5}$ particles for interior conditions without ventilation or forced ventilation is half that of the value in the exterior. In the case of natural ventilation, the difference between inside and outside is almost negligible.

In the analysis of the particle size 10.0 $\mu g/m^3$ (PM_{10}), there is no appreciable difference between concentrations indoors and outdoors in conditions without ventilation or with natural ventilation, as in both cases, the highest concentration is outside. Finally, in the case of forced ventilation, the concentration sampled outside is double that sampled inside since the filters retain the larger particles.

Geriatric centres are usually full, and the residents are typically sedentary people who move around within small spaces. Residents are also usually together for a long time in different areas (motor room, activity room, dining room, TV room, etc.), which have a high concentration of harmful agents and stale air that must be properly renewed. To optimise the indoor air for these people, good maintenance of air-conditioning equipment is necessary as well as an adequate change of filters.

Based on all of the above and taking into account the two ventilation systems used (natural and forced), it would be logical for nursing homes to use mechanical ventilation systems assisted by natural ventilation.

For the statistical analyses described below, given the condition of forced ventilation, the PM_{10} and $PM_{2.5}$ concentrations outside could potentially be regarded as being twice as high as those inside. However, these data were obtained in a passenger compartment (study area) without people. The areas studied in this work are geriatric centres, where there is a great deal of mass, so for this study, the concentrations of both particle types were considered to be identical outside and inside, using the data obtained in Table 2. This hypothesis is described in the Indoor Air Quality study [53] carried out for the National Institute of Safety and Health (INSST) in Spain.

A statistical analysis was performed with the support of the Statgraphics Centurion v.xvi program to determine the influence of all parameters on the number of elderly deaths. A principal component analysis was carried out in order to reduce the number of variables studied with respect to the number of elderly deaths in the homes of each AACC. An analysis of variance was then carried out to identify the convergence of the variables selected in the parameter studied.

3. Results

As indicated at the end of Section 1, the analysis was carried out in two stages, first taking into account the sizes of the different residences, and subsequently evaluating the effects of physical agents and pollutants.

For the first stage, the statistical technique of multivariate correlation between the variables was applied in order to determine the relationships between the number of deaths in the residences and the 12 variables related to the size of the residence. Then, a cluster or conglomerate analysis was carried out to determine the relationship between the nursing homes and the autonomous communities under study.

A total of 3844 private residences and 1573 public residences have been counted in Spain [31]. At this stage of the study, the nursing homes were categorised into private and public, and these, in turn, were classified by their size, resulting in 12 variables, including private residence with less than 25 people (PR < 25), private residence with between 25 and 49 people (PR25_49), private residence with between 50 and 99 people (PR50_99), residences with more than 100 people (PR > 100), residences with an unknown number of people (PR_NI) and the total number of private residences (PR). Similar variables were established for public residences for the elderly, resulting in the following variables: PU < 25, PU25_49, PU50_99, PU > 100, PU_NI and PU [31].

For the statistical study, a multivariate analysis was applied, where the number of elderly persons (old people) who have died in homes is a dependent variable, and the 12 previous variables are independent variables. The following table (Table 3) shows the Pearson correlation coefficients between each pair of variables. The correlation coefficient varies from -1 to $+1$ and shows the strength of the linear relationship between the variables.

The pairs of variables with P values below 0.05 (95% confidence level) were obtained from the analysis to identify statistically significant correlations. Since the aim is to assess the relationship between residences and the number of deaths in them, the relationships between the RD variable and the other 12 variables must be calculated to then identify independent variables that are significantly related to RD. The variables with the most significant relationships with RD were the following (the *p*-values obtained are in parentheses): PR25_49 (0.0153), PR50_99 (0.0020), PR $\geq$ 100 (0.0001), PU50_99 (0.0428) and PU $\geq$ 100 (0.0000).

It is also observed that for medium-sized and large nursing homes, size is related to the number of deaths; that is, a larger size is associated with more deaths. For different residence types (private or public) with the same size, for example, PR50_99 and PU50_99, the relationship between the number of deceased persons and the size is more significant for the private residence.

Table 3. Correlations between the different variables analysed.

	CC.AA.	(RD) Residential Deads	(RD) Community Deads	(TD) Total Deads	PR<25 Private Residences (<25)	PR25_49 (25–49)	PR50_93 (50–93)	PR ≥ 100 (≥100)	PR_NI No inf.	PR Total Privates:	PU < 25 Public Residences (<25)	PU25_49 (25–49)	PU50_93 (50–93)	PU ≥ 100 (≥100)	PU_NI No inf.	PU Total Publics:
CC.AA		0.0458	0.0706	0.0819	−0.0371	−0.2717	−0.2076	−0.1577	−0.4264	−0.2107	−0.1546	−0.3066	−0.278	−0.1034	−0.2121	−0.273
(RD)	0.0458		0.9699	0.9737	0.8801	0.5474	0.6624	0.7889	−0.0744	0.6228	0.2255	0.2891	0.4689	0.8157	−0.0005	0.4605
(CD)	0.0706	0.9699		0.9862	0.4415	0.5704	0.6802	0.8094	−0.0124	0.6595	0.1169	0.1946	0.5065	0.8079	0.0174	0.3872
(TD)	0.0819	0.9437	0.9862		0.545	0.6585	0.749	0.8319	−0.0041	0.7366	0.0492	0.2005	0.5835	0.7946	−0.042	0.3741
PR < 25	−0.0371	0.3659	0.4415	0.545		0.8066	0.7652	0.5907	0.1824	0.8462	0.02	0.3536	0.7683	0.6352	−0.0992	0.4313
PR25_49	−0.2719	0.5474	0.57	0.65	0.8		0.96	0.79	0.33	0.97	0.12	0.58	0.92	0.69	0.11	0.6
PR50_93	−0.2	0.66	0.68	0.74	0.76	0.96		0.87	0.23	0.97	0.22	0.57	0.92	0.69	0.11	0.6
PR ≥ 100	−0.15	0.78	0.8	0.83	0.59	0.79	0.82		0.35	0.87	0.14	0.48	0.71	0.86	0.28	0.55
PR_NI	−0.42	−0.074	−0.012	−0.004	0.18	0.33	0.23	0.35		0.34	−0.02	0.37	0.24	0.23	0.47	0.21
PR	−0.21	0.62	0.65	0.73	0.84	0.97	0.97	0.87	0.34		0.14	0.56	0.9	0.79	0.15	0.61
PU < 25	−0.15	0.22	0.11	0.04	0.02	0.12	0.22	0.14	−0.02	0.14		0.71	0.2	0.39	0.6	0.83
PU25_49	−0.3	0.28	0.19	0.2	0.35	0.58	0.57	0.48	0.37	0.56	0.71		0.62	0.55	0.58	0.92
PU50_93	−0.27	0.46	0.5	0.58	0.76	0.92	0.9	0.71	0.24	0.9	0.2	0.62		0.63	0.23	0.66
PU ≥ 100	−0.1	0.81	0.8	0.79	0.63	0.69	0.77	0.86	0.23	0.79	0.39	0.55	0.63		0.24	0.69
PU_NI	−0.21	−0.0005	0.0174	−0.0042	−0.00992	0.11	0.18	0.28	0.47	0.15	0.6	0.58	0.23	0.24		0.59
PU	−0.27	0.46	0.38	0.37	0.43	0.6	0.66	0.55	0.21	0.61	0.83	0.92	0.66	0.69	0.59	

Next, a cluster analysis was performed using the Ward method and the Euclidean square distance metric. Four groups or clusters were established from 19 observations in order to study the groupings of the five variables with the greatest significance or strongest relation with the number of elderly persons who have died in the residences. The six variables of greatest significance are RD, PR25_49, PR50_99, PR ≥ 100, PU50_93 and PU ≥ 100, which were analysed with respect to autonomous communities.

The results are expressed in a graph or dendrogram (Figure 3), which is a type of graphical representation or data diagram in the form of a tree that organises data into subcategories, which are further divided into other subgroups until the desired level of detail is reached. The dendrogram tool uses a hierarchical clustering algorithm. The program first calculates the distances between each pair of classes in the input signature file. It then iteratively merges the closest pair of classes and successively merges the next closest pair of classes and the next closest pair of classes until all classes are merged. After each merge, the distances between all class pairs are updated. The distances at which class signatures are merged are used to construct the dendrogram.

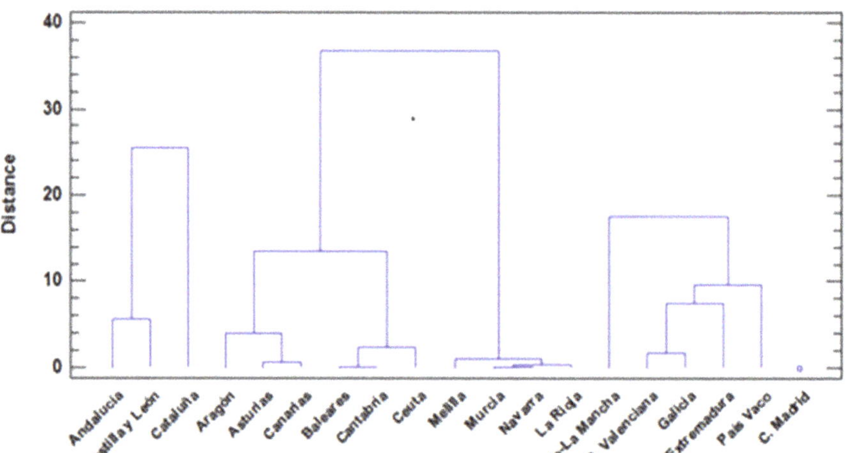

Figure 3. Dendrogram with the four final groups of analysis.

This graph (Figure 3) shows four groups with similar characteristics, according to the variables analysed. To form the groups, the procedure starts with each observation in separate groups, and then the two observations that are closest are combined to form a new group. After recalculating the distance between groups, the closest groups are combined again, and this process is repeated until all four groups are formed.

Figure 3 shows four very clearly defined groups. The first group (G1) is formed by the communities of Catalonia, Andalucia and Castilla y León, and the second group (G2) is formed by the communities of Castilla–La Mancha, C. Valenciana, Galicia, Extremadura and País Vasco. The third group (G3) contains only Madrid, which appears as a point at the lower right of the dendrogram. The rest of the autonomous communities form the fourth group (G4).

Relating the six previous variables of influence with the four associations or groups of autonomous communities, it was found that the variable most related to the number of deaths in all of the autonomous communities is PR $\geq$ 100, which denotes private residences with more than 100 persons (large old people's homes). This relationship is shown in the cluster dispersion diagram in Figure 4. Cluster analysis is a multivariate technique that allows cases or variables in a dataset to be grouped according to the similarity between them, i.e., cluster analysis is a multivariate technique whose main objective is to classify objects by forming groups (clusters) whose within-group homogeneity and between-group heterogeneity are both as high as possible. In each cluster in Figure 4, central points called centroids can be seen. The centroid of a cluster is defined as the equidistant point of the objects belonging to that cluster.

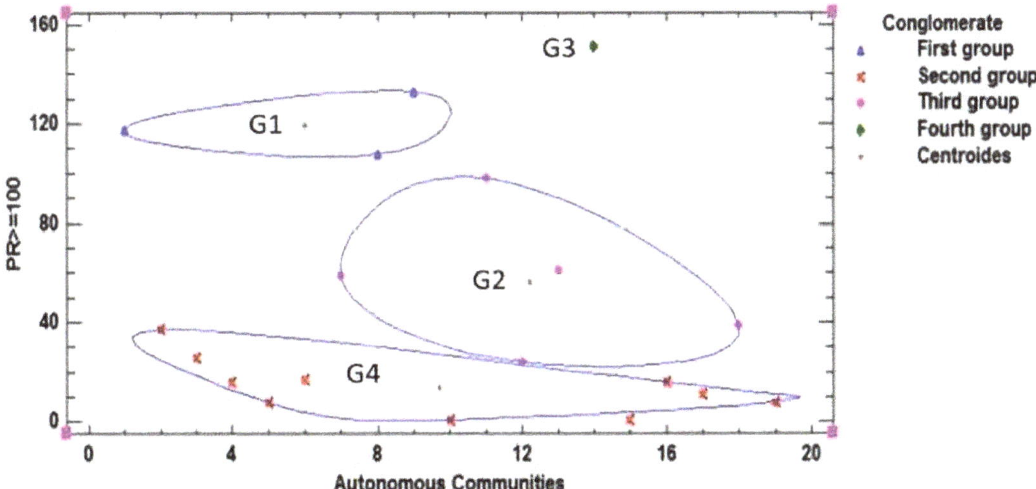

Figure 4. Diagram of the dispersion of conglomerates of the autonomous communities, compared to public residences with a capacity of more than 100 people.

For the second stage, the data corresponding to physical agents and chemical pollutants were analysed to identify relationships between the indoor air quality in the residences and the number of elderly people who have died in different residences in different autonomous communities of Spain. The analysis process was similar to the one carried out in the first stage. To analyse the influence of the size of the residences, a cluster or classification analysis was carried out together with other techniques to relate and compare variables.

The studied variables are the following physical agents: average temperature (AT) in degrees Celsius, average humidity (AH) as a percentage, precipitation (PREC) in millimetres, and hours of sunshine (SH) in hours, among which we included the population density of the communities (PD) in inhabitants per square kilometre. The following variables were the analysed pollutants: average annual particulate matter (PM_{10AA} and $PM_{2.5AA}$) measured in micrograms per cubic metre, average annual nitrogen dioxide (NO_2) measured in micrograms per cubic metre, average annual ozone (O3OCTNOR) measured in micrograms per cubic metre, average annual sulphur dioxide (SO_2) measured in micrograms per cubic metre and formaldehyde (FOR) in micrograms per square metre. In total, five variables belonging to physical agents (including population density) and six variables belonging to pollutants were included in this analysis.

The cluster analysis led to the following groups of autonomous communities according to the previous 11 variables and the number of elderly people who have died in the homes.

Figure 5 shows four different groups were obtained, which are labelled differently from those obtained in the analysis of the first stage in order to differentiate them: the first group (FC1) is formed by the communities of Andalusia, Castilla–La Mancha, Catalonia and the Canary Islands; the second group (FC2) is formed by Asturias, Cantabria, Extremadura and Melilla; the third group (FC3) contains only Galicia; and the fourth group (FC4) includes the rest of the autonomous communities.

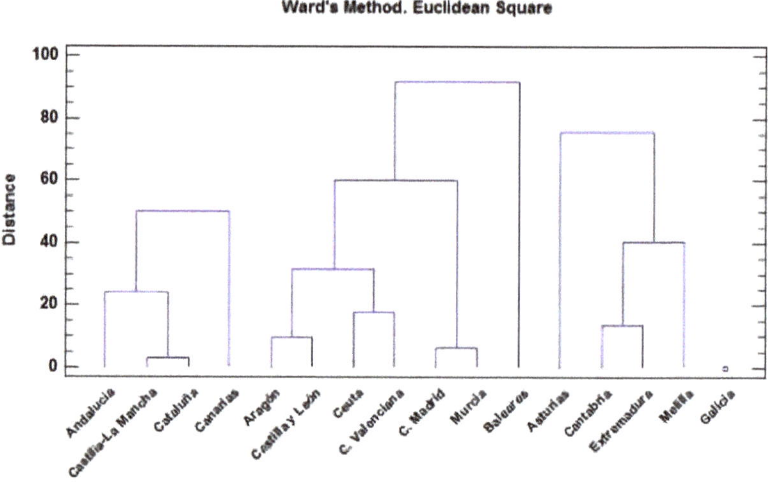

Figure 5. Cluster diagram of physical agents and contaminants in AACCs.

A multivariate factor analysis was performed using listwise deletion. Factorisation was applied for standardisation of the main components, and a varimax rotation was used to analyse these variables and identify possible relationships. The objective was to extract a small number of the 13 factors (the previous 11, plus DR and AACC) that explain most of the variability considered in the analysis.

Applying the previous statistical methodology, four factors were extracted, because these factors have eigenvalues greater than or equal to the unit. These four factors explain 79.3414% of the variability in all of the data which are shown in Table 4. When carrying out principal component analysis, the initial estimate of the community was established to assume that all variability in the data is due to common factors.

Table 4. Factor analysis table.

Factor		Variance	Cumulative
Number	Eigenvalor	Porcentage	Porcentage
1	4.09141	31.472	31.472
2	2.43407	18.724	50.196
3	2.28212	17.555	67.751
4	1.50677	11.591	−79.341
5	0.9121	7.016	86.358
6	0.67221	5.171	91.528
7	0.347443	2.673	94..201
8	0.315038	2.423	96.624
9	0..22916	1.763	98.387
10	0.099754	0.767	99.154
11	0.0636375	0.490	99.644
12	0.029118	0.224	99.868
13	0.0171641	0.132	100.000

A factor loading matrix was then generated after the Varimax rotation. This is carried out so that the factors have a few high loadings and many near-zero loadings on the variables. This means that there are factors with high correlations with a small number of variables and zero correlation with the rest, thus redistributing the variance of factors.

The following table (Table 5) shows the equations that estimate the values of the common factors after the rotation. The rotation was performed to simplify the explanation of the factors. The first rotated factor is the following equation:

$$\begin{aligned} F1 = {} & 0.251871(AA.CC.) - 0.347442(RD) - 0.555084(AT) + 0.912213(AH) + \\ & 0.927481(PREC) - 0.920873(SH) - 0.0857457(DP) - 0.070322(PM10AA) + \\ & 0.0782349(PM2{,}5AA) + 0.383983(NO2) - 0.704236(O3OCTNOR) + 0.326536(SO2) + \\ & 0.324599(FOR) \end{aligned} \tag{1}$$

Table 5. Factor loading matrix after applying Varimax rotation.

	Factor	Factor	Factor	Factor
	1	2	3	4
Autonomous Communities	0.251871	−0.471631	0.262884	−0.52807
RD	−0.347442	−0.0841992	0.744084	−0.367933
AT	−0.555084	0.1I2131	0.0783305	0.662564
AH	0.912213	0.0516618	−0.145363	0.185372
PREC	0.927481	−0.00501128	0.0524455	−0.0907723
SH	−0.920873	−0.145316	−0.0741857	0.215049
DP	−0.0857457	−0.09622712	0.886502	0.076693
PM10 AA	−0.070322	0.904561	0.0244	0.1497t4
PM 2.5 AA	0.0782149	0.714746	0.174668	0.128887
NO2	0.383983	0.424818	0.760336	0.172825
O3OCTNOR	−0.704236	−0.157771	0.455243	−0.23757
SO2	0.326536	0.72773	−0.200042	−0.13505
FOR	0.324599	−0.325889	0.0236463	0.75947

The values of the variables in the equation are standardised by subtracting their means and dividing by their standard deviations. The equations for the other three factors, F2, F3 and F4, are obtained in a similar way by simply substituting the coefficients in the corresponding column for each factor.

Next, three of the factors obtained are plotted in 3D to evaluate the contributions of the different variables to each factor.

To identify the parameters that influence factor 1, we must look at their position along the length of the box (from the front): the most important parameters are on the right, and the less important ones are on the left. For factor 2, we view the box from the right side (width) and use the same criteria as for factor 1. For factor 3, we examine the height of the box, with the most important ones toward the top and the less important ones toward the bottom.

As indicated by the Figure 6, the highest loadings for factor 1 are provided by the variables PREC, AH, FOR and NO$_2$, whereas AT, RD, SH and O3OCTNOR appear to have low loadings for the same factor. For factor 2, the variables PM$_{10AA}$, PM$_{2.5AA}$ and SO$_2$ appear to be very important contributors, and the AACC variable appears to be less important. For factor 3, DP, RD and NO$_2$ appear to be influential, and no candidate variables appear to be the contrary. For factor 4, a parallel analysis was carried out, and the results suggest that the variables that most contribute to this factor are FOR and AT, and RD and AACC are of little importance.

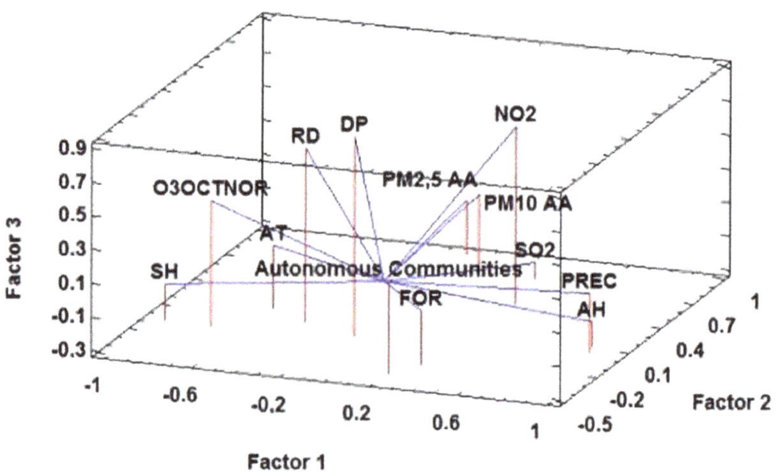

Figure 6. Graph of loadings for the different factors.

Multivariate analysis provides a series of Pearson correlation coefficients between each pair of variables as indicated above. For the two variables X and Y, this coefficient is defined as the quotient between the covariance of the two variables and the product of their standard deviations. The values of the linear correlation coefficient r range between -1 and $+1$: r = 0 indicates no correlation, r = $+1$ indicates the maximum positive correlation and linear dependence, and r = -1 indicates the maximum negative correlation and linear dependence. Values close to zero indicate a weak correlation, and values close to extreme values, either -1 or $+1$, indicate a strong correlation. The sign indicates whether the correlation is direct (positive) or inverse (negative). When the sign is positive, both variables vary in the same direction, either increasing or decreasing. When the sign is negative, even when there is a very high correlation between the two variables, it means that when one increases, the other decreases, i.e., they covary inversely.

The following relationships were found.

The Table 6 shows the pairs of variables identified in the analysis with P-values below 0.05, i.e., with a 95% confidence level. For a better interpretation of the analysis results, a bubble diagram was generated to illustrate the connections between variables. Two groups were obtained. In the first one, the connection of AH, AT, O3OCTNOR and SH to PREC stands out; that is, there is a high correlation between precipitation and humidity, temperature, ozone and sunshine hours. RD, DP and NO2 are connected to this first group through the ozone variable (O3OCTNOR). The main link in the chain, the number of elderly people who have died in homes (RD), depends on population density (PD) and ozone (O3OCTNOR), which in turn depend on the whole chain formed by the other variables analysed. The second group consists of particulate matter and sulphur dioxide. The PM_{10AA} particulate matter acts as a link between the other type of particulate matter ($PM_{2.5AA}$) and sulphur dioxide (SO_2). It is important to note that the formaldehyde variable (FOR) does not appear in any of the above combinations, which may indicate that its contribution to indoor air quality pollution in nursing homes is not very significant.

Table 6. Significant relationships between variables and bubble diagram.

Combination of Factors	Bubble Diagram
RD + DP	
RD + O3OCTNOR	
AH + PREC	
AH + SH	
AH + O3OCTNOR	
AT + PREC	
AT + SH	
PREC + SH	
PREC + O3OCTNOR	
PREC + NO2	
PM10AA + PM2.5AA	
PM10AA + SO2	

Carrying out a multivariate cluster or conglomerate analysis, the best combination obtained for the AACC, separated into four groups with all possible combinations with the other 12 variables under study, is the combination obtained with respect to the precipitation variable (PREC). Total significance is not achieved, as can be seen in the Figure 7, because two groups clearly intersect.

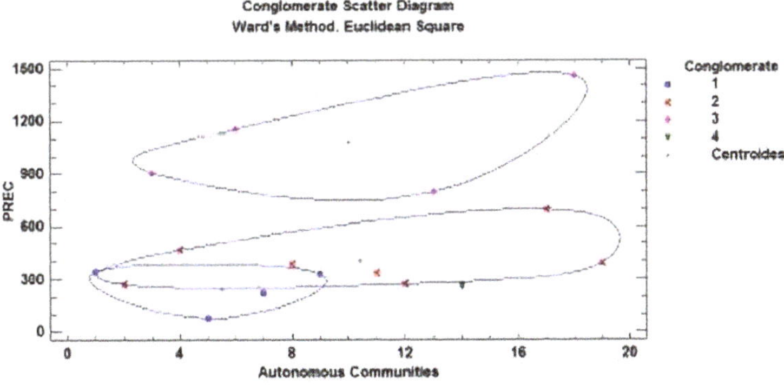

Figure 7. Greatest significance of all parameters against AACC.

All possible combinations between all variables were tested for a total of 156 combinations. The most significant cases of dependence are between the variables humidity (AH), nitrogen dioxide (NO_2), precipitation (PREC) and particulate matter (PM_{10AA}), as can be seen in the following graphs (Figure 8). The graph on the upper left shows the highest significance between the autonomous communities and all of the variables, although the relationship between the AACC and the humidity (HA) is not very strong, as there are intersections between the groups.

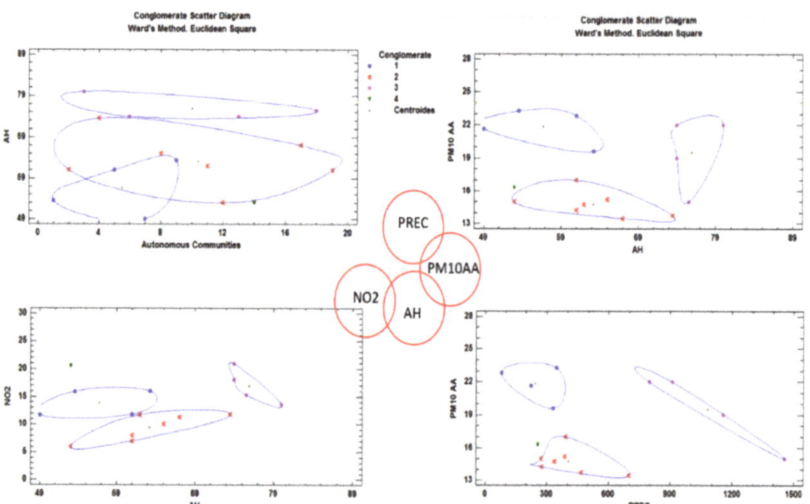

Figure 8. Most significant relationships among all parameters studied.

In the centre of Figure 8, the relationships through which these four variables influence the number of deaths in nursing homes are indicated by means of red circles.

4. Discussion

The analysis carried out in this work indicates that geriatric homes or senior centres with a capacity of over 100 people in the residence have had the highest number of deaths. The number of elderly persons who have died is greater in private homes than in public homes, probably due to a greater shortage of health resources and personnel, due to their cost.

It was possible to divide both private and public residences, together with the number of deaths therein, into four groups with common or similar characteristics. The first group is formed by the communities of Cataluña, Andalucía and Castilla y León. The second group is formed by Castilla–La Mancha, C. Valenciana, Galicia, Extremadura and País Vasco. The third group is formed only by the Community of Madrid, and finally, the fourth group is formed by the rest of the communities.

Establishing a link between atmospheric factors and harmful effects on human health poses some difficulties. The effects on health are highly variable, and each factor can generate more than one effect on human beings. The physical agents studied, in addition to population density, have a clear connection with the number of deaths through ozone (Table 6). The main meteorological factors that have been shown to have a clear effect on human health are air temperature, humidity, wind speed, hours of solar radiation, pressure and precipitation [54,55], many of which were also analysed in this work, and their dependency relationship was demonstrated.

If the relationship between the number of elderly deaths and the total number of deaths in each community is taken into account, the communities of Castilla–La Mancha (14.40%) and Extremadura (14.13%) present the most significant results. If we study the environmental conditions (physical agents) in these communities compared to their average values in Spain (AT = 16.6 °C, AH = 60.4%, SH = 2739.8 h and PREC = 496.7 mm), then possible contributors to the higher rates of elderly deaths in these regions emerge: their relative humidity is higher than the average, temperatures and hours of sunshine are approximately equal to the average values for the entirety of Spain and, finally, precipitation is lower than the average. Excessive humidity may favour the proliferation of moulds and mites, which increase the risk of respiratory infections [56]. Health problems related to the presence of heat waves (lower rainfall) have worsened, affecting the most vulnerable population groups, i.e., the elderly and children, to a greater extent. The communities of Extremadura and Castilla–La Mancha have a Mediterranean climate type characterised by hot, dry summers with little rainfall in the summer period and long, milder winters.

Six pollutants were studied: ozone, particulate matter ($PM_{2.5}$ and PM_{10}), sulphur dioxide, nitrogen dioxide and formaldehyde. The results of the analysis indicate that the effects of formaldehyde were not significant. The contribution of this agent to indoor air quality pollution due to construction materials and furniture in the residence was estimated at 3–4% [53].

The ozone values in the communities of Extremadura and Castilla–La Mancha are 32% higher than the average for Spain (60.4 $\mu g/m^3$). Some studies have shown a direct relationship between daily mortality and tropospheric ozone levels [57]. As tropospheric ozone concentrations increase above the recommended limit (100 $\mu g/m^3$), health effects are exacerbated, so prolonged exposure can have chronic effects on the respiratory system. These effects may become more pronounced in very hot weather.

The relationship between the two tested types of particulate matter (PM_{10} and $PM_{2.5}$) and sulphur dioxide has high significance (Table 6). Anthropogenic activity is the main factor influencing air pollution levels on hourly and daily time scales. However, a clear relationship has also been found between African desert dust intrusions (Saharan air layer) and the number of times that the daily PM_{10} limit value is exceeded. The contribution of these long-range particulate transport processes to the annual average levels of PM has been especially relevant in the southern and central regions of the peninsula, Melilla and the Canary and Balearic archipelagos. The average annual contribution of African dust episodes to PM_{10} and $PM_{2.5}$ levels has also been quantified. There is also a large variation between the north of the peninsula, with annual contributions of 1–2 $\mu gPM10/m^3$ and <1 $\mu gPM_{2.5}/m^3$ to the Canary Islands, Melilla and the Balearic Islands, and the south of the peninsula, with 4–6 $\mu gPM_{10}/m^3$ and 1–2 $\mu gPM_{2.5}/m^3$ [58].

The factors with the greatest influence on the temporal variability of PM concentration levels include locally recorded surges of anthropogenic pollution (mainly traffic emissions, with some industrial exceptions, demolition–construction, domestic and residential emissions), followed by increases in natural and anthropogenic pollution produced on a regional scale and by African dust episodes. The highest rates of incident solar radiation recorded in the summer months favour the formation of secondary particles, as well as the re-suspension of mineral dust by convective processes in semi-arid environments in the peninsula, such as the autonomous communities of Andalucia, Extremadura and Castilla–La Mancha, which have average $PM_{10}/PM_{2.5}$ values of (23.27/13.27), (21.66/10.66) and (15/8) in micrograms per cubic metre, respectively. As can be seen from the data, the community closest to the emission of particulate matter from the Sahara Desert has the highest values of particulate matter, compared to the community of Castilla–La Mancha, which is the farthest away, but all of them are strongly affected by this material.

In indoor environments, sulphur dioxide levels are much lower than those outside, provided that there are no indoor sources that might produce it, such as kerosene stoves, boilers or chimneys. For this reason, it is not a pollutant that causes major problems indoors. However, the existence of particles produces a synergistic effect in the presence of sulphur dioxide, since the combination of these two substances produces a greater effect than each substance alone. Sulphur oxides also contribute to the formation of acid rain, as do nitrogen oxides. Murcia and Asturias are the communities with the highest rates of this pollutant. For the communities with the highest rate of elderly deaths in the residence/community, this parameter affects very little. The highest incidence occurs in the community of Extremadura.

The statistical combination of the population density (PD) variable shows that it has the lowest correlation or significance with all of the variables analysed, so it can be ruled out as having an influence on the number of elderly persons who died in the homes.

One more relation that was obtained is the one among precipitation (PREC), humidity (AH), nitrogen dioxide (NO_2) and particulate matter (PM_{10AA}). The concentration of PM_{10} is strongly related to relative humidity and precipitation (Figure 8, upper right and lower right, respectively). The higher the precipitation, the higher the humidity and the higher the particulate matter concentration [59,60]. On the other hand, relative humidity (RH) is closely related to nitrogen oxide (Figure 8, bottom left). There are several internal sources of nitrogen oxide emissions. On the one hand, those that occur in outdoor environments can infiltrate into indoor environments through air change processes. On the other hand, combustion processes in indoor environments are the main internal source of nitrogen oxide emissions. The decreases in industrial, transport and commercial activity since the beginning of the coronavirus outbreak have considerably reduced the levels of atmospheric nitrogen dioxide (NO_2) in the AACC, so it is considered a minor factor in this analysis. However, the increase in humidity also causes an increase in the concentration of nitrogen dioxide. For the communities of Extremadura and Castilla–La Mancha, NO_2 concentrations show values of 10.66 $\mu g/m^3$ and 8 $\mu g/m^3$, respectively, below the national average of 13.89 $\mu g/m^3$; these values are within the regulations.

Based on the above discussion, it is difficult to establish precisely to what extent indoor air quality may affect the health of elderly people in residential homes, as the data related to exposure and the effect of concentrations are not sufficient. On the one hand, although the effects of acute exposure to many air pollutants are well known, there are important gaps in the data concerning long-term exposures to low concentrations and mixtures of different pollutants. On the other hand, only experience and the rational design of ventilation, occupancy and residential compartmentalisation will guarantee better indoor air quality.

In the last 20 years, studies on the presence of pollutants in many indoor environments have shown higher levels than expected. In addition, pollutants different from those present in outdoor air have been identified. The presence of microorganisms in indoor air can lead to problems of an infectious and allergic nature. Viruses, as in the case of COVID-19, can cause acute respiratory diseases, especially in the most vulnerable people, such as the elderly, particularly those in residences, where the occupation density is usually high and there is significant air recirculation [60–63].

The ventilation of a building, and thus the ventilation of a nursing home, is measured in air changes per hour. This means that every hour, a volume of air equal to the volume of the nursing home enters from the outside; similarly, every hour, a similar volume of air is expelled from the inside. If there is no forced ventilation, this value usually varies between 0.2 and 2 changes per hour. The concentration of pollutants produced inside the residence will be lower with high values of air changes, although this is not a guarantee of air quality. The compromise between air quality and energy efficiency is of great importance, especially when the residences are private, where the cost factor is very important.

Efforts to reduce the air velocity inside the building, as well as to increase the insulation and waterproofing of residences, require the installation of materials that can be sources of indoor air pollution.

In recent decades, the trend towards designing buildings, including nursing homes, with minimal energy use has led to the development of buildings with very low air infiltration and exfiltration, implemented through equipment such as HVAC systems, which has generally enabled higher concentrations of airborne microorganisms and other types of pollutants. The placement of air inlets near cooling towers or other sources of microorganisms, in addition to making it difficult to access the HVAC system for maintenance and cleaning, is a flaw in the design, operation and maintenance that can affect the health of the occupants, especially the elderly, who live in the building.

5. Conclusions

Based on the data obtained, greater attention should be paid to private residences with a capacity of over 100 people, where the number of deaths is very high.

The data show that the physical agents studied, in addition to meteorological factors, have a clear connection with the number of deaths.

The autonomous communities of Extremadura and Castilla–La Mancha are the most affected by a high number of COVID-19 deaths in the most vulnerable population groups.

Of the pollutants studied, formaldehyde has the least influence on air quality pollution (3–4%).

In addition to carbon dioxide, PM_{10} and $PM_{2.5}$ have a strong influence in the southern regions of Spain, Melilla and the Canary and Balearic archipelagos due to dried dust emissions from Africa.

The population density has the weakest correlation with the other variables, so it may be possible to omit this variable in subsequent studies.

PM_{10} concentration is strongly related to relative humidity and precipitation, and relative humidity is strongly related to nitrogen oxide.

Indoor combustion processes are the main source of nitrogen oxide emissions. Increased humidity also leads to increased nitrogen dioxide concentrations.

Special care must be taken with the data on pollutant concentrations in indoor environments, which, unfortunately, are often below the actual levels.

It is advisable to increase the number of air changes in residences to improve air quality, even if this results in higher energy costs. Health should be prioritised over cost.

These issues should be addressed in the design of buildings, particularly those of nursing homes, where significant deficiencies have been shown in the design of heating, air-conditioning and ventilation systems.

Authors should discuss the results and how they can be interpreted from the perspective of previous studies and of the working hypotheses. The findings and their implications should be discussed in the broadest context possible. Future research directions may also be highlighted.

Author Contributions: Conceptualization, G.S.-C., E.J.L.-F. and R.A.G.-L.; methodology, G.S.-C., E.J.L.-F. and R.A.G.-L.; software, G.S.-C., E.J.L.-F. and R.A.G.-L.; validation, G.S.-C., E.J.L.-F. and R.A.G.-L.; formal analysis G.S.-C., E.J.L.-F. and R.A.G.-L.; investigation, G.S.-C., E.J.L.-F. and R.A.G.-L.; resources G.S.-C., E.J.L.-F. and R.A.G.-L.; data curation, G.S.-C., E.J.L.-F. and R.A.G.-L.; writing—original draft preparation, G.S.-C., E.J.L.-F. and R.A.G.-L.; writing—review and editing, G.S.-C., E.J.L.-F. and R.A.G.-L.; visualization, G.S.-C., E.J.L.-F. and R.A.G.-L.; supervision, G.S.-C. and R.A.G.-L. All authors have read and agreed to the published version of the manuscript.

Funding: This research received no external funding.

Conflicts of Interest: None of the authors have conflict of interest associated with this study to report.

References

1. Klepeis, N.E.; Nelson, W.C.; Ott, W.R.; Robinson, J.P.; Tsang, A.M.; Switzer, P.; Behar, J.V.; Hern, S.C.; Engelmann, W.H. The national human activity pattern survey (NHAPS): A resource for assessing exposure to environmental pollutants. *J. Expo. Sci. Environ. Epidemiol.* **2001**, *11*, 231–252. [CrossRef] [PubMed]
2. Wargocki, P. Ventilation, Indoor Air Quality, Health, and Productivity. In *Ergonomic Workplace Design for Health, Wellness, and Productivity*; Wargocki, P., Ed.; CRC Press: Boca Raton, FL, USA, 2016; pp. 39–72.
3. Borsboom, W.; De Gids, W.; Logue, J.; Sherman, M.; Wargocki, P.; Civil Engineering, Technical University of Denmark (Eds.) *Technical Note AIVC 68 Residential Ventilation and Health*; INIVE: Brussels, Belgium, 2016.
4. Wargocki, P. The effects of outdoor air supply rate in an office on perceived air quality, sick building syndrome (SBS) symptoms and productivity. *Indoor Air* **2000**, *10*, 222–236. [CrossRef]
5. Nathanson, T. *Indoor Air Quality in Office Buildings: A Technical Guide*; Health Canadá: Ottawa, ON, Canada, 1993.
6. Cao, G.; Awbi, H.; Yao, R.; Fan, Y.; Sirén, K.; Kosonen, R.; Zhang, J. A review of the performance of different ventilation and airflow distribution systems in buildings. *Build. Environ.* **2014**, *73*, 171–186. [CrossRef]
7. Li, Y.; Leung, G.M.; Tang, J.W.; Yang, X.; Chao CY, H.; Lin, J.Z.; Yuen, P.L. Role of ventilation in airborne transmission of infectious agents in the built environment-A multidisciplinary systematic review. *Indoor Air* **2007**, *17*, 2–18. [CrossRef] [PubMed]
8. Daisey, J.M.; Angell, W.J.; Apte, M.G. Indoor air quality, ventilation and health symptoms in schools: An analysis of existing information. *Indoor Air* **2003**, *13*, 53–64. [CrossRef]
9. Wagner, H.U.; Verhoeff, A.P.; Colombi, A.; Flannigan, B.; Gravesen, S.; Moulliseux, A.; Nevelainen, A.; Papadakis, J.; Sebel, K. Biological Particles. In *In-Door Environments. Indoor Air Quality and Its Impact on Man*; Comisión Europea: Brussels, Belgium, 1993.

10. Wargocki, P. *What Are Indoor Air Quality Priorities for Energy-Efficient Buildings? Indoor and Built Environment*; SAGE Publications Ltd.: Thousand Oaks, CA, USA, 2015. [CrossRef]

11. González-Lezcano, R.A.; Hormigos-Jiménez, S. Energy saving Due to Natural Ventilation in Housing Blocks in Madrid. In *IOP Conference Series: Materials Science and Engineering*; Institute of Physics Publishing: Bristol, UK, 2016; Volume 138. [CrossRef]

12. Organización Mundial de la Salud. *Air Quality Guidelines for Europe, European Series, no. 23*; Publicaciones Regionales de la OMS: Copenhagen, Denmark, 1987.

13. Yocom, J.E.; McCarthy, S.M. *Measuring Indoor Air Quality: A Practical Guide*; Wiley: Chichester, UK, 1991.

14. Spengler, J.D.; Chen, Q. Indoor air quality factors in designing a healthy building. *Annu. Rev. Energy Environ.* **2000**, *25*, 567–600. [CrossRef]

15. Hormigos-Jimenez, S.; Padilla-Marcos, M.A.; Meiss, A.; Gonzalez-Lezcano, R.A.; Feijó-Muñoz, J. Experimental validation of the age-of-the-air CFD analysis: A case study. *Sci. Technol. Built Environ.* **2018**, *24*, 994–1003. [CrossRef]

16. Turiel, I. *Indoor Air Quality and Human Health*; Stanford University Press: Palo Alto, CA, USA, 1986.

17. Berenguer, M.J.; Guardino, X.; Hernández, A.; Martí, M.C.; Nogareda, C.; Solé, M.D. *El Síndrome del Edificio Enfermo. Guía Para su Evaluación*; Instituto Nacional de Seguridad e Higiene en el Trabajo: Madrid, Spain, 1994.

18. Wadden, R.A.; Schelf, P.A. *Indoor Air Pollution. Characterization, Prediction and Control*; Wiley: New York, NY, USA, 1983.

19. Dimitroulopoulou, C. Ventilation in European dwellings: A review. *Build. Environ.* **2012**, *47*, 109–125. [CrossRef]

20. Van Buggenhout, S.; Zerihun Desta, T.; Van Brecht, A.; Vranken, E.; Quanten, S.; Van Malcot, W.; Berckmans, D. Data-based mechanistic modelling approach to determine the age of air in a ventilated space. *Build. Environ.* **2006**, *41*, 557–567. [CrossRef]

21. Xu, C.; Luo, X.; Yu, C.; Cao, S.J. *The 2019-nCoV Epidemic Control Strategies and Future Challenges of Building Healthy Smart Cities. Indoor and Built Environment*; SAGE Publications Ltd.: Thousand Oaks, CA, USA, 2020. [CrossRef]

22. Código Técnico de la Edificación (CTE). *Documento Básico HS3: Calidad del Aire Interior*; Ministerio de Fomento del Gobierno de España: Madrid, Spain, 2018.

23. Fernández, M. *Sistema Integral de Ventilación de Viviendas de Acuerdo con el Código Técnico de la Edificación (HS3). Instalaciones y Técnicas del Confort*; Ministerio De Fomento: Madrid, Spain, 2008; pp. 10–21.

24. RITE. *Reglamento de Instalaciones Térmicas de los Edificios*; Ministerio para la Transición Ecológica y el Reto Demográfico: Madrid, Spain, 2020.

25. Namiesnik, J.; Gorecki, T.; Krosdon-Zabiegala, B.; Lusiak, J. Indoor air quality pollutants, their sources and concentration levels. *Build Environ.* **1992**, *27*, 339–356. [CrossRef]

26. Knoppel, H.; Wolkoff, P. *Chemical, Microbiological, Health and Confort Aspects of Indoor Air Quality-State of the Art in SBS*; Kluwert Academic: Dordrecht, The Netherlands, 1992.

27. Davidson, L.; Olsson, E. Calculation of age and local purging flow rate in rooms. *Build. Environ.* **1987**, *22*, 111–127. [CrossRef]

28. Kwon, K.S.; Lee, I.B.; Han, H.T.; Shin, C.Y.; Hwang, H.S.; Hong, S.W.; Han, C.P. Analysing ventilation efficiency in a test chamber using age-of-air concept and CFD technology. *Biosyst. Eng.* **2011**, *110*, 421–433. [CrossRef]

29. Li, J.; Tartarini, F. Changes in Air Quality during the COVID-19 Lockdown in Singapore and Associations with Human Mobility Trends. *Aerosol Air Qual. Res.* **2020**, *20*, 1748–1758. [CrossRef]

30. Liu, M.; Tilton, J.N. Spatial distributions of mean age and higher moments in steady continuous flows. *AIChE J.* **2010**, *56*, 2561–2572. [CrossRef]

31. CCAyES. Centro de Coordinación de Alertas y Emergencias Sanitarias del Ministerio de Sanidad. 2020. Available online: https://www.mscbs.gob.es/profesionales/saludPublica/ccayes/home.htm (accessed on 6 March 2021).

32. Abellán García, A.; Aceituno Nieto, P.; Ramiro Fariñas, D. Estadísticas Sobre Residencias: Distribución de Centros y Plazas Residenciales por Provincia. Datos de Abril de 2019. Available online: http://envejecimiento.csic.es/documentos/documentos/enred-estadisticasresidencias2019.pdf (accessed on 22 October 2021).

33. INE 2020. Instituto Nacional de Estadística. Available online: https://www.ine.es/covid/covid_inicio.htm (accessed on 6 March 2021).

34. Rasmussen, M.; Foldbjerg, P.; Christoffersen, J.; Daniell, J.; Bang, U.; Galiatto, N.; Bjerre, K. *Healthy Homes Barometer 2017-Buildings and Their Impact on the Health of Europeans*; Rasmussen, M.K., Ed.; VELUX Group: Hørsholm, Denmark, 2017.

35. Sanglier, G.; Robas, M.; Jiménez, P.A. Gamma Radiation in Aid of the Population in Covid-19 Type Pandemics. *Contemporary Eng. Sci.* **2020**, *13*, 113–129. [CrossRef]

36. Wargocki, P. The Effects of Ventilation in Homes on Health. *Int. J. Vent.* **2013**, *12*, 101–118. [CrossRef]

37. Kim, S.; Kim, J.A.; Kim, H.J.; Do Kim, S. Determination of formaldehyde and TVOC emission factor from wood-based composites by small chamber method. *Polym. Test.* **2006**, *25*, 605–614. [CrossRef]

38. Guo, H. Source apportionment of volatile organic compounds in Hong Kong homes. *Build. Environ.* **2011**, *46*, 2280–2286. [CrossRef]

39. World Health Organization. *Guidelines for Indoor Air Quality: Selected Pollutants*; WHO Regional Office for Europe: Bonn, Germany, 2010.

40. *Prevención de la Exposición a Formaldehído*; Nota Técnica de Prevención; Instituto Nacional de Seguridad e Higiene en el Trabajo (INSHT): Madrid, Spain, 2010; Volume 830.

41. Stymne, H.; Axel Boman, C.; Kronvall, J. Measuring ventilation rates in the Swedish housing stock. *Build. Environ.* **1994**, *29*, 373–379. [CrossRef]

42. Kostiainen, R. Volatile organic compounds in the indoor air of normal and sick houses. *Atmos. Environ.* **1995**, *29*, 693–702. [CrossRef]
43. Tham, K.W. Indoor air quality and its effects on humans—A review of challenges and developments in the last 30 years. *Energy Build.* **2016**, *130*, 637–650. [CrossRef]
44. European Collaborative Action. *Indoor Air Quality and its Impact on Man. Report No.11: Guidelines for Ventilation Requirements in Buildings*; Office fot Publications of EU: Luxembourg, 1992.
45. UNE EN ISO 7730:2006. *Determinación analítica e interpretación del bienestar térmico mediante el cálculo de los índices PMV y PPD y los criterios de bienestar térmico local*; ISO: Geneva, Switzerland, 2006.
46. AEMet. Agencia Estatal de Meteorología. 2020. Available online: http://www.aemet.es/es/portada (accessed on 12 February 2021).
47. Health Canadá. 2005. Available online: https://www.canada.ca/en/health-canada/services/health-care-system/reports-publications/canada-health-act-annual-reports/annual-report-2004-2005.html (accessed on 6 January 2021).
48. Directiva 2008/50/CE Del Parlamento Europeo y del Consejo. 2008. Relativa a la calidad del aire ambiente y a una atmósfera más limpia en Europa. Available online: https://www.boe.es/doue/2008/152/L00001-00044.pdf (accessed on 14 March 2021).
49. Real Decreto 102/2011 (2011). Relativo a la mejora de la calidad del aire. Available online: https://www.boe.es/buscar/act.php?id=BOE-A-2011-1645 (accessed on 10 December 2020).
50. Organización Mundial de la Salud (OMS) y Organización Meteorológica Mundial. *Atlas de la Salud y del Clima*; World Health Organization: Geneva, Switzerland, 2012.
51. Ceballos, M.A.; Segura, P.; Gutiérrez, E.; Gracia, J.C.; Ramos, P.; Reaño, M.; García, B. *La Calidad del Aire en el Estado Español Durante 2018*; Ecologistas en Acción: Madrid, Spain, 2018.
52. Hormigos-Jimenez, S.; Padilla-Marcos, M.Á.; Meiss, A.; Gonzalez-Lezcano, R.A.; Feijó-Muñoz, J. Computational fluid dynamics evaluation of the furniture arrangement for ventilation efficiency. *Build. Serv. Eng. Res. Technol.* **2018**, *39*, 557–571. [CrossRef]
53. Saéz, E. *Análisis de la Calidad del Aire en Función de la Tipología de Ventilación. Aplicación al Frototipo E3. Edificación Eco-Eficiente de la UPM*; Universidad Politécnica de Valencia: Valencia, Spain, 2017.
54. Guardino, X. *Calidad del Aire Interior*; Instituto Nacional de Seguridad e Higiene en el Trabajo (INSHT): Madrid, Spain, 2012; Chapter 44; p. 36.
55. Sanglier, G.; Robas, M.; Jiménez, P.A. Application of Forecasting to a Mathematical Models for adjusting the Determination of the Number of People Infected by Covid-19 in the Comunity of Madrid (Spain). *Contemp. Eng. Sci.* **2020**, *13*, 89–101. [CrossRef]
56. Fernández-García, J.; Dosil Díaz, O.; Taboada Hidalgo, J.; Fernández, J.; Sánchez-Santos, L. Influencia del clima en el infarto de miocardio en Galicia. *Med. Clín.* **2014**, *145*, 97–101. [CrossRef]
57. Monsalve, F. *Influencia del Tiempo y de la Contaminación Atmosférica Sbre Enfermedades de los Sistemas Circulatorio y Respiratorio en Castilla-La Mancha*; University of La Rioja: La Rioja, Spain, 2011.
58. Organización Mundial de la Salud (OMS/WHO). *Guías de Calidad del Aire de la OMS Relativas al Material Particulado, el Ozono, el Dióxido de Nitrógeno y el Dióxido de Azufre*; World Health Organization: Geneva, Switzerland, 2005.
59. Querol, X.; Viana, M.; Moreno, T.; Alastuy, A. *Bases Científicos-Técnicas Para un Plan Nacional de Mejora de la Calidad del Aire*; Spanish National Research Council: Madrid, Spain, 2012.
60. Sanglier, G.; González, R.A.; Cesteros, S.; López, E.J. Validation of the Mathematical Model Applied to Four Autonomous Communities in Spain to Determine the Number of People Infected by Covid-19. *Mod. Appl. Sci.* **2020**, *14*. [CrossRef]
61. Villalba, D.; Fajardo, E.; Romero, H. Relación entre el material particulado PM10 y variables meteorológicas en la ciudad de Bucaramanga–Colombia: Una aplicación del análisis de datos longitudinal. In Proceedings of the XXVIII Simposio Internacional de Estadística, Bucaramanga, Colombia, 23–27 July 2018.
62. Harapan Harapan Itoh, N.; Yufita, A.; Winardi, W.; Te, H.; Megawati, D.; Hayati, Z.; Wagner, A.L.; Mudatsir, M. Coronavirus disease 2019 (COVID-19): A literature review. *J. Infect. Public Health* **2020**, *13*, 667–673. [CrossRef]
63. Law, R.C.K.; Lai, J.H.K.; Edwards, J.D.; Hou, H. COVID-19: Research directions for non-clinical aerosol-generating facilities in the built environmental. *Buildings* **2021**, *11*, 282. [CrossRef]

 *sustainability*

Article

The Effect of Basalt Fiber on Mechanical, Microstructural, and High-Temperature Properties of Fly Ash-Based and Basalt Powder Waste-Filled Sustainable Geopolymer Mortar

Mahmoud Ziada [1], Savaş Erdem [1,*], Yosra Tammam [2], Serenay Kara [1] and Roberto Alonso González Lezcano [3,*]

[1] School of Civil Engineering, Avcilar Campus, Istanbul University-Cerrahpasa, 34200 Istanbul, Turkey; m.ziada@ogr.iu.edu.tr (M.Z.); serenaykara1@gmail.com (S.K.)

[2] Civil Engineering Department, Avcilar Campus, Istanbul Gelisim University, 34200 Istanbul, Turkey; yosra.tammam@ogr.iu.edu.tr

[3] Architecture and Design Department, Escuela Politécnica Superior, Universidad CEU San Pablo, 28040 Madrid, Spain

* Correspondence: savas.erdem@iuc.edu.tr (S.E.); rgonzalezcano@ceu.es (R.A.G.L.)

Abstract: As the human population grows and technology advances, the demand for concrete and cement grows. However, it is critical to propose alternative ecologically suitable options to cement, the primary binder in concrete. Numerous researchers have recently concentrated their efforts on geopolymer mortars to accomplish this objective. The effects of basalt fiber (BF) on a geopolymer based on fly ash (FA) and basalt powder waste (BP) filled were studied in this research. The compressive and flexural strength, Charpy impact, and capillary water absorption tests were performed on produced samples after 28 days. Then, produced samples were exposed to the high-temperature test. Weight change, flexural strength, compressive strength, UPV, and microstructural tests of the specimens were performed after and before the effect of the high temperature. In addition, the results tests conducted on the specimens were compared after and before the high-temperature test. The findings indicated that BF had beneficial benefits, mainly when 1.2 percent BF was used. When the findings of samples containing 1.2 percent BF exposed to various temperatures were analyzed, it was revealed that it could increase compressive strength by up to 18 percent and flexural strength by up to 44 percent. In this study, the addition of BF to fly ash-based geopolymer samples improved the high-temperature resistance and mechanical properties.

Keywords: geopolymer; fly ash; basalt fiber; basalt waste aggregate; mechanical properties

Citation: Ziada, M.; Erdem, S.; Tammam, Y.; Kara, S.; Lezcano, R.A.G. The Effect of Basalt Fiber on Mechanical, Microstructural, and High-Temperature Properties of Fly Ash-Based and Basalt Powder Waste-Filled Sustainable Geopolymer Mortar. *Sustainability* **2021**, *13*, 12610. https://doi.org/10.3390/su132212610

Academic Editor: Jose Ignacio Alvarez

Received: 26 October 2021
Accepted: 12 November 2021
Published: 15 November 2021

Publisher's Note: MDPI stays neutral with regard to jurisdictional claims in published maps and institutional affiliations.

1. Introduction

In terms of environmental protection, it is critical to creating a sustainable product. The geopolymer mortars were used to reduce OPC use to as an alternative to Ordinary Portland cement (OPC). Cement manufacture accounts for 5–7 percent of global CO_2 emissions [1]. This scenario has severe environmental consequences. Finding alternate OPC replacements is essential. Industrial waste materials (slag, red mud, fly ash) were significant replacements. Alkaline activators were used to activate these waste materials in the geopolymer manufacture process. The carbon dioxide emissions are decreased due to recycling waste materials and using less OPC [2].

Additionally, geopolymer mortars have much superior durability and mechanical properties. In this instance, the compact and dense geopolymer microstructure is beneficial [3]. Industrial waste, produced as a byproduct of many technological operations, is a major environmental issue. One of these sectors is the quarry industry which generates a large quantity of industrial waste. The environmental cost for storing waste items rises year after year. The mining and processing of rock raw materials are rapidly expanding. Rock raw materials are treated to cutting and grinding operations in stone manufacturers. The fine-grained waste material generated during processing is transported to and deposited

119

in landfills. Therefore, geopolymer was produced using different waste materials, and its properties were investigated in previous studies [4–7]. The sodium hydroxide (SH) and sodium silicate (SS) used in this study were close to the amounts of SH and SS used in these studies. However, it can be stated that using BP instead of sand in this study is more economical than geopolymers using sand. BP is a byproduct of the manufacturing of high-quality aggregates from basalt rocks. Tammam et al. [8] were used industrial waste filler materials such as BP, marble powder, and limestone powder waste in geopolymer mortars. They found that the use of industrial waste enhanced the mechanical and durability properties of the geopolymer specimens. Celikten and Atabey [9] investigated the mechanical and physical characteristics of geopolymer mortars manufactured using basalt stone cutting waste. Geopolymer mortars with four different water contents were created for this purpose. They found that the mortars with the least water content had the highest strength values.

Improving mechanical characteristics using fiber has significant benefits. PVA, basalt, glass, Carbon, and steel fiber have all been used to create geopolymer [10–12]. Previous research used polyvinyl alcohol (PVA) fibers and nano-SiO_2 to manufacture geopolymer concrete. The findings showed that PVA fibers enhanced the proprieties' strength, improved toughness, and achieved the optimal value at 0.6% to 1% [13–16]. Significant values were achieved with high temperatures in BF, and PVA contained geopolymer mortars. BF was examined and is among the most widely used concrete fibers [17]. It is a very efficient insulating, non-toxic, low-cost fibrous material formed during basalt rock melting at up to 1400 °C. It is less costly than other fibers since it requires less energy and no additives [18]. Additionally, it was shown that it had a function in strength growth by substantially adding to workability. Several publications have verified the beneficial benefits of BF [19–23]. Ali et al. [24] investigated the impact of boron waste and BF on metakaolin-based geopolymer. Based on the findings of their study, 24 mm BF has more significant beneficial impacts than 12 mm BF. The specimens were found to be stable during high temperatures, sulfuric acid, and freeze-thaw effects.

Previous studies were conducted on alternative binder ingredients in fly ash-based geopolymer. Only a few of them have mixed various ratios of BF and basalt waste aggregate. This study aims to produce a sustainable mortar with high mechanical and microstructural properties and good high-temperature resistance. Thus, this study has led to environmentally friendly geopolymer mortar production using waste materials instead of cement and sand. For this proposal, BF contained fly ash-based, and BP-filled geopolymer mortar samples were produced. The mechanical, physical, and microstructural tests were performed after and before the high-temperature test.

2. Materials and Methods

2.1. Materials

In this study, waste materials were used as binder and filler material in fly ash-based geopolymer mortars. Class F fly ash (FA) and slag (S) were used as binders. Basalt dust waste (BP) was also used as filling material. The physical and chemical properties of BP, S, and FA materials are given in Table 1. The particle diameters of BP range from 1 to 300 μm, with an average particle diameter of 120 μm. Sodium hydroxide (SH) and sodium silicate (SS) were used as activators to produce geopolymer mortar. Tables 2 and 3 illustrate the chemical characteristics of SS and SH, respectively. In addition, BF was used at 4, 8, and 12 percent by volume. BF properties used in blends are shown in Table 4. The images of FA, S, BP, and BF materials used in fly ash -based and BP-filled geopolymer mortars are given in Figure 1.

Table 1. The physical and chemical properties of FA, S, and BP materials.

Materials	SiO$_2$	Al$_2$O$_3$	Fe$_2$O$_3$	TiO$_2$	CaO	MnO	MgO	K$_2$O	Na$_2$O	SO$_3$	Cl-	Loss of Ignition	Specific Gravity	Blaine (cm^2/g)
FA (%)	54.08	26.08	6.681	-	2.002	-	2.676	-	0.79	0.735	0.092	1.36	1.98	2571
BP (%)	56.9	17.6	8.1	0.9	8.15	0.1	2.1	1.9	3.8	-	-	-	2.76	6285
S (%)	40.55	12.83	1.1	-	35.58	-	5.87	-	0.79	0.18	0.0143	0.03	2.9	2612

Table 2. The chemical properties of SS (wt. %).

Na$_2$O (%)	SiO$_2$ (%)	Density (20 °C) (g/mL)	Fe (%)	Heavy Metals Value (pb) %
9.68	26.12	1.367	<0.005	<0.005

Table 3. The chemical properties of SH (%).

NaOH (g/kg)	Na$_2$CO$_3$ (g/kg)	SO$_4$	Fe	Cl	Al
$\geq$990	$\leq$4	$\leq$0.01	$\leq$0.002	$\leq$0.01	$\leq$0.002

Table 4. Basalt fiber properties.

Diameter (mm)	Length (mm)	Elasticity Module (GPa)	Nominal tensile Strength (MPa)	Specific Gravity
0.02	12	88	4100	2.73

Figure 1. The materials used in geopolymer mortars: (**a**) Fly ash; (**b**) slag; (**c**) basalt powder; and (**d**) basalt fiber.

2.2. Mixing Procedure and Samples

In the first mixing stage of the fly ash-based geopolymer mortars produced in this study, 12 M sodium hydroxide solution was prepared and kept at 20 °C for 24 h to cool.

1000 g of deionized water was used for each 40 g of sodium hydroxide pellets to prepare 12 M sodium hydroxide solution. Then, sodium hydroxide solution (SH) was mixed with sodium silicate solution (SS) after cooling. Then, fly ash was placed in the mixer, and the prepared mixture of SS and SH (mixed activator solution) was added and mixed until homogeneous. Then, slag was added to the mixture and mixed homogeneously. BP waste, which is the filling material, was added to the mixture and mixed homogeneously. Finally, BF was added to the mixtures at different rates (0%, 4%, 8%, and 12%) and mixed homogeneously. In addition, the amounts of the materials used in the fly ash-based geopolymer mortar mixtures produced in this study are given in (kg/m^3), as Table 5 shows. BF was added at 0%, 4%, 8%, and 12% by volume. Previous trial experiments and earlier studies were used to prepare the mixture [8,25].

Table 5. Mix design of fly ash-based geopolymer mortar series (kg/m^3).

Series	FA	Slag	SS	SH	BP	BF (Volume %)
0 BF	550	71	261	128	1100	-
4 BF	550	71	261	128	1100	4
8 BF	550	71	261	128	1100	8
12 BF	550	71	261	128	1100	1.2

Freshly prepared mixtures were placed into $40 \times 40 \times 160$ mm^3 prismatic molds and $50 \times 50 \times 50$ mm^3 cubic molds. The molds were then vibrated while being placed on the vibration table. Then the samples were cured in an oven at 60 °C for 24 h. The cured samples were kept at 20 °C for 28 days.

2.3. Tests Conducted

In this study, mixtures of fly ash-based geopolymer contained BF were prepared. Compressive strength, flexural strength, Ultrasonic pulse velocity (UPV), Charpy impact, and capillary water absorption tests were performed to examine the impact of BF content on prepared samples. Then, $50 \times 50 \times 50$ mm^3 cubic and $40 \times 40 \times 160$ mm^3 prismatic samples were subjected to the high-temperature test. Weight change, compressive strength, flexural strength, and UPV tests of the samples exposed to the high-temperature test were performed. In addition, the results before and after the high-temperature test were compared. Scanning Electron Microscopy (SEM) and Energy Dispersive X-ray (EDS) analyzes were performed to observe the BF content and changes in the matrices of the samples.

2.3.1. Compressive Strength, Flexural Strength, and Ultrasonic Pulse Velocity (UPV) Tests

Compressive strength, flexural strength, and UPV tests were performed to examine the impact of BF on produced samples after 28 days at 20 °C. The same tests were also carried out on samples exposed to high-temperature effects. For each mixture, the average compressive strength of three $50 \times 50 \times 50$ mm^3 cube samples after 28 days was performed. Compressive strength was conducted with a rate of 0.602 MPa/s according to ASTM C39. In addition, compressive strengths were performed on the geopolymer samples exposed to 200 °C, 400 °C, 600 °C, and 800 °C temperatures, and the differences between the results were analyzed. The flexural test was performed after 28 days on three $40 \times 40 \times 160$ mm^3 prism samples for each mixture. Flexural strength tests were conducted following ASTM C 348. As in the compressive strength test, the flexural strength test was also performed after the high-temperature test. The $4\times$ and $40\times$ magnified crack images of 12 BF specimens subjected to flexural strength test are shown in Figure 2.

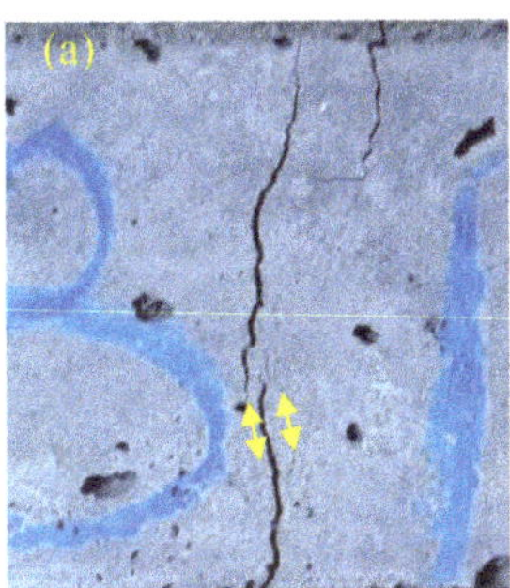
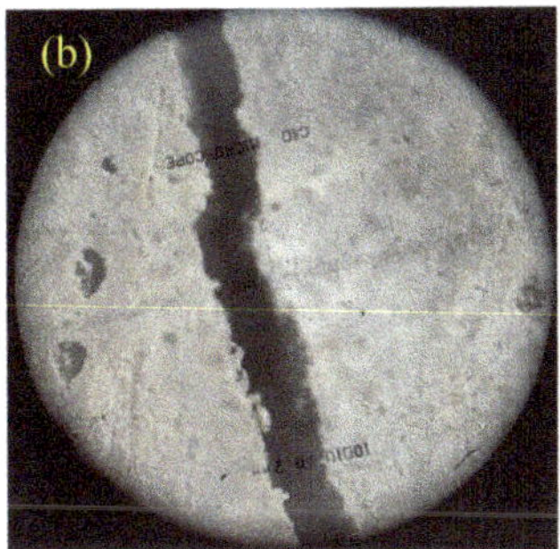

Figure 2. Crack images of 12 BF specimens subjected to flexural strength test magnified at (**a**) 4×; (**b**) 40×.

In addition, the UPV test of the geopolymer mortar samples was applied after 28 days according to ASTM C597-16. UPV test results of fly ash-based geopolymer samples contained BF exposed to 20 °C, and high temperatures were obtained. The damage degrees caused by high temperatures were calculated using the UPV results and the following formula:

$$Dd = 1 - \left(\frac{v_a}{v_b} \right) \tag{1}$$

Dd was the damage degree of samples, V_a was the UPV values of samples after the high-temperature test, and V_b was the UPV values of samples before the high-temperature test.

2.3.2. Charpy Impact Test

The Charpy impact test was conducted following ASTM E23. $40 \times 40 \times 160$ mm^3 prismatic fly ash-based geopolymer specimens were used for the Charpy impact test. In the first stage of the Charpy impact test, the sample was placed centered on the device, as shown in Figure 3a. Subsequently, the pendulum was raised, and the drag indicator was set to the zero position. Finally, the pendulum was released from the first height and swung through the specimen to the second height. Figure 3 illustrates the procedure of the Charpy impact test and the samples subjected to this test. Three fly ash-based geopolymer samples containing BF were subjected to the Charpy impact test, and the average results were recorded from the device in the Kgf-m unit.

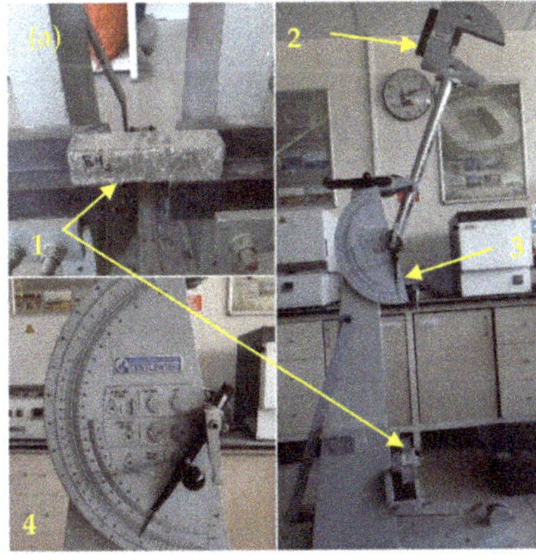

Figure 3. (**a**) Charpy impact test procedure; (**b**) Samples exposed to Charpy impact test.

2.3.3. Capillary Water Absorption

The capillary water absorption capacity of fly ash-based geopolymer samples was determined using ASTM C1585-20. $40 \times 40 \times 160$ mm^3 fly ash-based geopolymer samples were dried in an oven at 105 °C for 24 h. Then, the samples surfaces were covered with paraffin except for the surface contacted with water and the opposite surface. Then, the dry weights (Wd) of the samples were obtained, and the dry samples were immersed in 5 mm depth water. After 15 min, 30 min, 1, 2, 3, 5, 24, 48, 72, and 96 h, the fly ash-based samples were withdrawn from the water, and their wet weight (Ww) was determined. Finally, the capillary water absorption curves and water absorption coefficient (*Kc*) using the following formula:

$$K_C \times \sqrt{t} = \frac{Q}{A} \tag{2}$$

where Q was the absorbed water mass (Ww-Wd) (kg), Wd was the dry weight (kg), Ww was the wet weight (kg), A was the immersed surface area (m^2), t was a time (s), and Kc was the water absorption coefficient $((kg/(m^2. \sqrt{s}))$.

2.3.4. High-Temperature Test

In this study, a high-temperature test was performed to examine the high-temperature resistance of fly ash-based geopolymer samples and the effect on the BF contained in the samples. Fly ash-based geopolymer samples were dried in an oven at 105 °C for 24 h. Dried samples were exposed to temperatures of 200 °C, 400 °C, 600 °C, and 800 °C, respectively. The high-temperature degrees rose with the rate of 5 °C/min, and then the samples were exposed to temperatures for another 1 h.

The dry weight (Wd) of the samples dried at 105 °C for 24 h in the oven was performed. Then, the weights of the samples exposed to the effects of high temperatures were measured, and the percentages of change in weights were calculated. The images of the cracks were obtained using a microscope with $40\times$ magnification to evaluate the effect of high temperature on the samples.

3. Results and Discussions

3.1. Compressive and Flexural Strength Results

Compressive strength, flexural strength, UPV, and Charpy impact tests were performed on geopolymer mortar samples after 28 days. These tests were conducted to study the effect of BF and determine the mechanical properties of the fly ash-based geopolymer mortars filled with BP. Subsequently, the results of these tests were determined. Figures 4 and 5 show the compressive and flexural strength results of fly ash-based geopolymer samples containing BF. Additionally, Figures 4 and 5 show the relationship between the strengths and the content of BF in the geopolymer series. The relationships were quite strong, and R^2 was 0.94 for compressive strength (Figure 4) and 0.97 for flexural strength (Figure 5). Therefore, in this study, it was found that the increase of BF content in fly ash-based geopolymer mortars increased the compressive and flexural strength of the specimens. The increases in compressive strength were 6.43%, 9.65%, and 11.94% for 4 BF, 8 BF, and 12 BF, respectively.

Similarly, the flexural strength increased by 7.72%, 25.51%, and 34.15% for 4 BF, 8 BF, and 12 BF, respectively. Consequently, the increase in flexural strength due to basalt content was higher than the increase in compressive strength. Especially for the 12 BF series, the increase in flexural strength was 2.86 times larger than the increase in compressive strength. All specimens with BF exhibited higher strength than the 0 BF specimens. The use of BF increased the compressive and strength of the geopolymer matrix. The delicate and uniform distribution of crystal phases helped improve the BF materials' physical and mechanical properties. Applying a nucleating agent such as P_2O_5, TiO_2, and ZrO_2 helped achieve the required degree of microstructure. At the same time, basaltic rock was not required for melting, creating a naturally occurring nucleating agent such as Fe_3O_4 aided in the process. Thus, the nucleating agents outperformed the necessary fibers [17].

The addition of BF significantly reduced the number of fractures during the flexural test because they transferred the flexural load. In other words, they delayed the formation of cracks by distributing the resulting stress. As a result, BF had a massive effect on flexural strength [26].

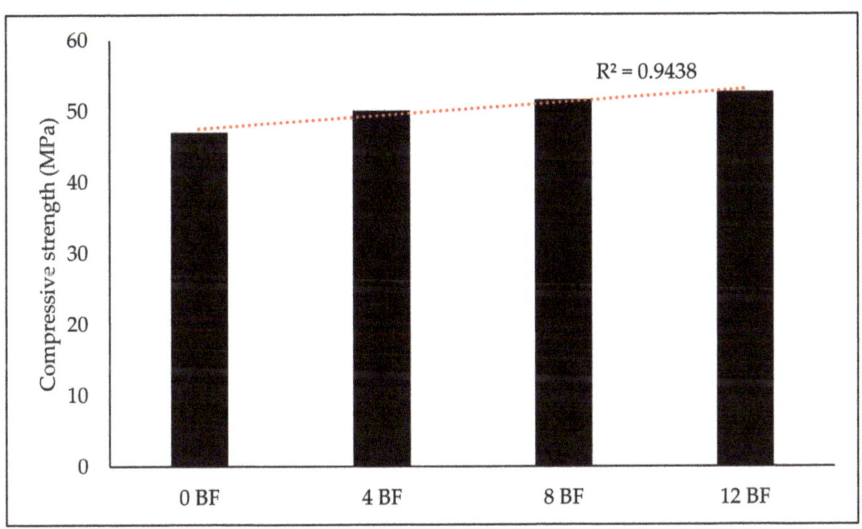

Figure 4. The relationship between the compressive strength and the content of BF in geopolymer series.

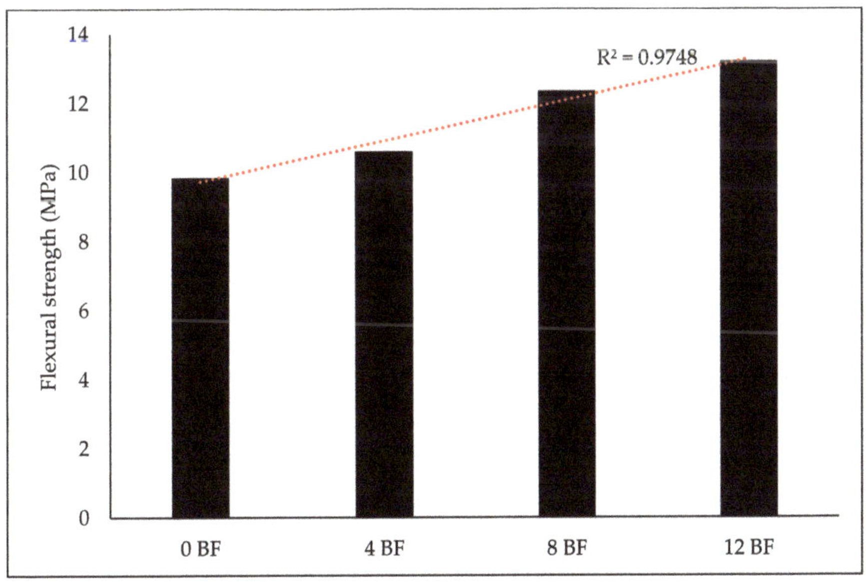

Figure 5. The relationship between the flexural strength and the content of BF in geopolymer series.

3.2. Ultrasonic Pulse Velocity (UPV) Results

The UPV test was performed on fly ash-based geopolymer mortar samples after 28 days. The UPV test was carried out to study the effect of BF on the samples and to determine the compactness of the fly ash-based geopolymer mortars filled with BP waste. Figure 6 shows the UPV test results of the series. Similarly, Figure 6 illustrates the relationship between the UPV and BF content in the geopolymer series. The wave velocity and the number of voids in the matrix of geopolymer samples have a reciprocal relationship [27]. The penetration rate is determined by dividing the distance between the origin and the receiver by the elapsed time. When the matrix of samples contains more voids, the time required increases, and the penetration rate decreases.

Consequently, the void ratio has a significant effect on the mechanical properties of fly ash-based geopolymer specimens. The relationship between UPV and BF content was relatively high, with an R^2 of 0.99. Thus, in this study, it was found that the addition of BF to fly ash-based geopolymer mortars increased the UPV values of the geopolymer specimens. The UPV values increased by 3.34%, 6.05%, and 8.90% for series 4 BF, 8 BF, and 12 BF, respectively, compared to series 0 BF. Although the fiber exposure improved the UPV values by reducing the void ratio, the improvement was not significant. This situation was because the fibers had no direct effect on the homogeneity or compaction of the matrix. The observed behavior indicates that basalt fiber content influences the UPV of composites and is connected to the composite density and perceived porosity. Additionally, basalt fiber influences the geopolymer matrix and improved the UPV values of geopolymer mortar samples. Ozkan and Coban [28] investigated the hybrid effects of basalt and PVA fiber on the properties of cementitious composites, and the effect of BF on UPV results they found are consistent with the effect of BF performed in this study.

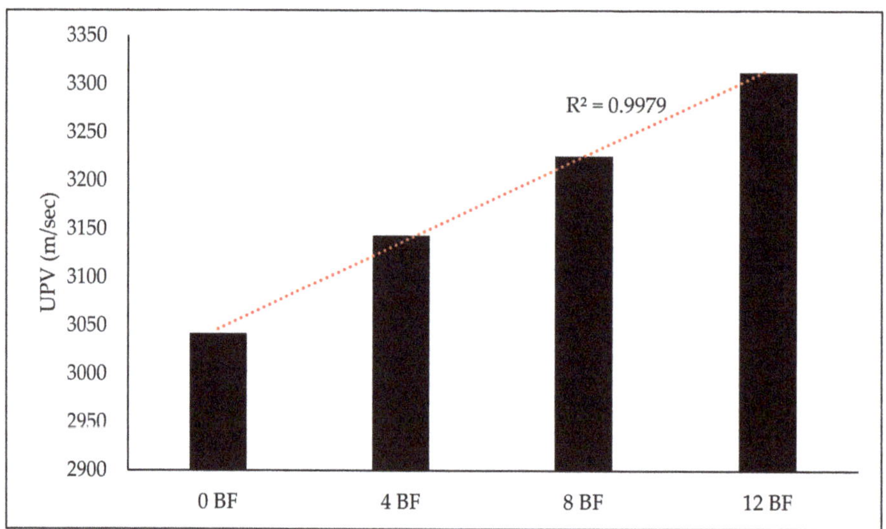

Figure 6. The relationship between the UPV and the content of BF in geopolymer series.

3.3. Charpy Impact Test Results

The Charpy impact test was performed on prismatic fly ash-based geopolymer samples. The Charpy impact test was performed on samples of the series, and the average results were reported in Kgf-m units. This test uses a falling pendulum's potential energy to fracture samples at high strain rates and characterize the composites. Mechanical property testing procedures are based on the principles of stress wave propagation, kinetic energy, or potential energy [29]. The absorption of impact energy and comparable characteristics are highly dependent on the fiber properties [30]. In this study, the correlation between Charpy impact test values and BF content was strong, and the R2 was 0.90. The addition of BF enhanced the Charpy impact test values of the geopolymer specimens. The Charpy impact test results were 13.65, 24.4, 28.65, and 29.8 Kgf-m for series 0 BF, 4 BF, 8 BF, and 12 BF, respectively. The Charpy impact test results increased by 78.75%, 109.89%, and 118.32% for series 4 BF, 8 BF, and 12 BF, respectively, compared to series 0 BF. The Charpy impact test results increases are shown in Figure 7.

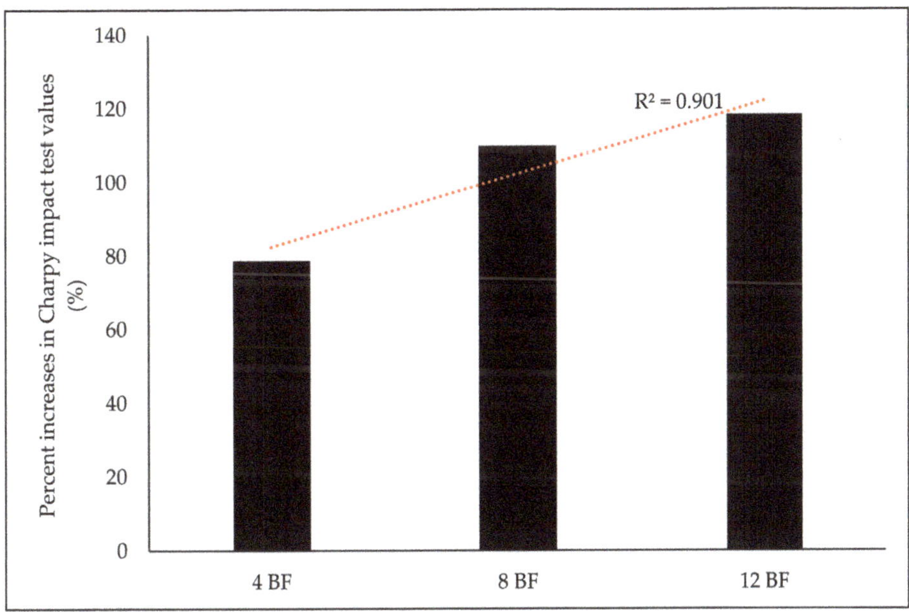

Figure 7. The increases of Charpy impact test values of fly ash-based geopolymers.

3.4. Water Absorption

After 28 days, the capillary water absorption test was performed on the samples, and the results are shown in Figure 8. In this study, the amount of water absorbed over time was determined as a function of the area in contact with the water, and the capillary water absorption curves shown in Figure 8 were obtained using formula two given in Section 2.3.3. Series 0 BF absorbed the least amount of water, and series 12 BF absorbed the most water compared to the other series. Increasing the BF percentage increased the amount of water absorbed, as shown by the slopes of the capillary water uptake. The percentage increase in water absorption was 9.38%, 11.88%, and 18.13% for 4 BF, 8 BF, and 12 BF, respectively, compared to sample 0 BF. These results are supported by the slopes of the capillary water uptake tests. Figure 9 also shows the capillary water absorption coefficient (Kc) values of fly ash-based geopolymer mortar samples containing BF in different proportions. The water absorption coefficient (Kc) values were determined using the least-squares method. The percentage increases in the water absorption coefficients of samples 4 BF, 8 BF, and 12 BF were 9.04%, 10.55%, and 17.99%, respectively. As a result of capillary water absorption of fly ash geopolymer samples containing BF in different ratios, increasing the content of BF increased the amount of water absorbed by the samples. This increase is because the addition of BF improves pore connectivity at large volume fractions. Similar results were obtained in [21,31].

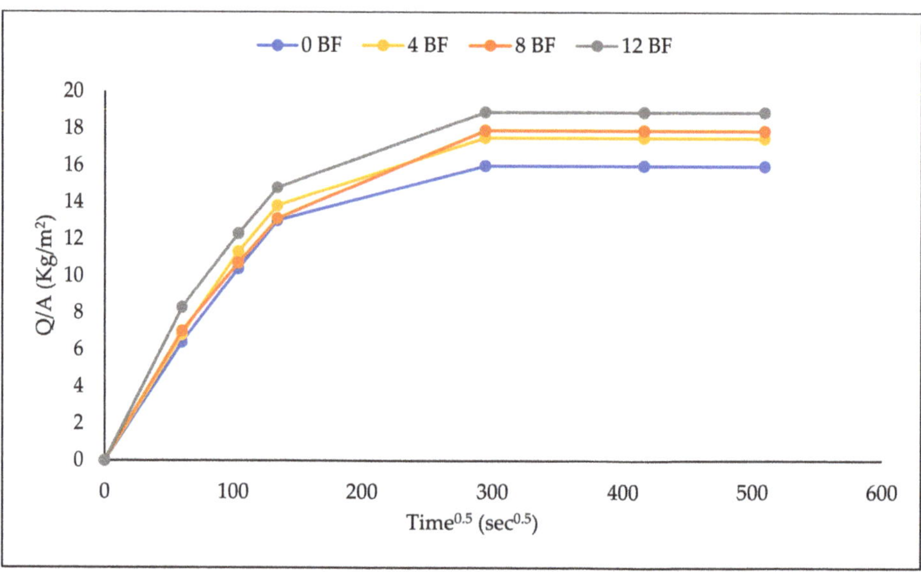

Figure 8. The amount of water absorbed per m^2 of fly ash-based geopolymer mortar samples contained BF with different ratios.

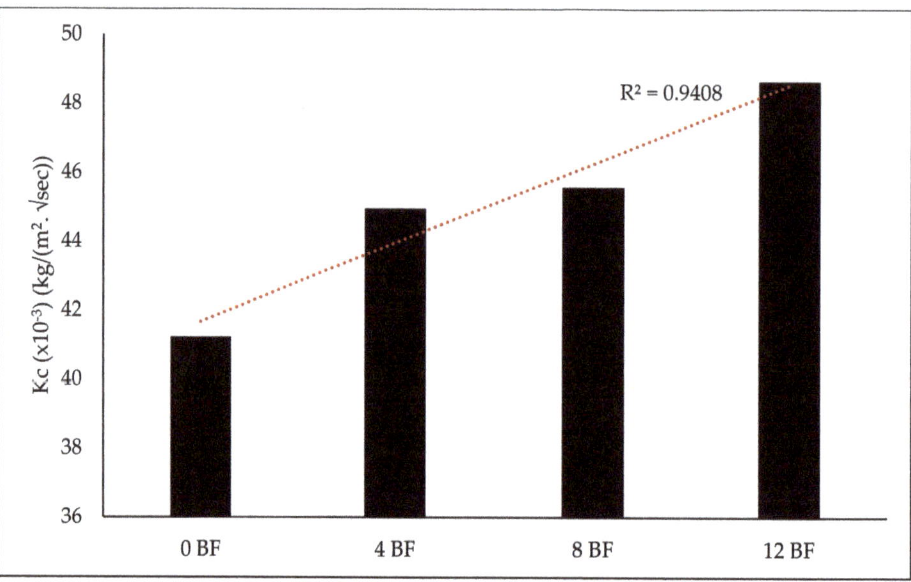

Figure 9. Fly ash-based geopolymer mortar samples' capillary water absorption coefficient (*Kc*) contained BF with different ratios.

3.5. High-Temperature Results

3.5.1. Weight Measurement Results

The behaviors of the produced geopolymer specimens after a high-temperature test were analyzed and compared with samples cured at 20 °C for 28 days. Fly ash-based samples were exposed to 200 °C, 400 °C, 600 °C, and 800 °C temperatures to perform the high-temperature test. Figures 10–15 showed the weight loss, compressive strength, flexural strength, UPV, and damage degree values of fly ash geopolymer samples after and before the high-temperature test. After 400 °C, the increase in weight loss ratio was observed, as seen in Figure 10. The 12 BF samples had the smallest weight loss, and the 0 BF had the highest weight loss. After 800 °C, the weight loss percentages of 0 BF, 4 BF, 8 BF,12 BF were 8.73%, 8.49%, 8.38%, and 8.17% respectively. The inclusion of BF slowed down the weight reduction. This occurred as a result of BF's ability to limit the high-temperature cracks in the samples. Thus, reducing the formation of cracks reduced the samples' weight loss exposed to the high-temperature effect. Sahin et al. [32] investigated the effect of basalt fiber on metakaolin-based geopolymer mortars, and they found that the inclusion of BF reduced weight reduction. Generally, the interface bonds between paste and aggregate weaken at high temperatures, leading to weight loss [33,34]. For example, the weight loss percentages of 12 BF samples exposed to 200 °C, 400 °C, 600 °C, and 800 °C were 1.26%, 4.26%, 7.18%, and 8.17 respectively.

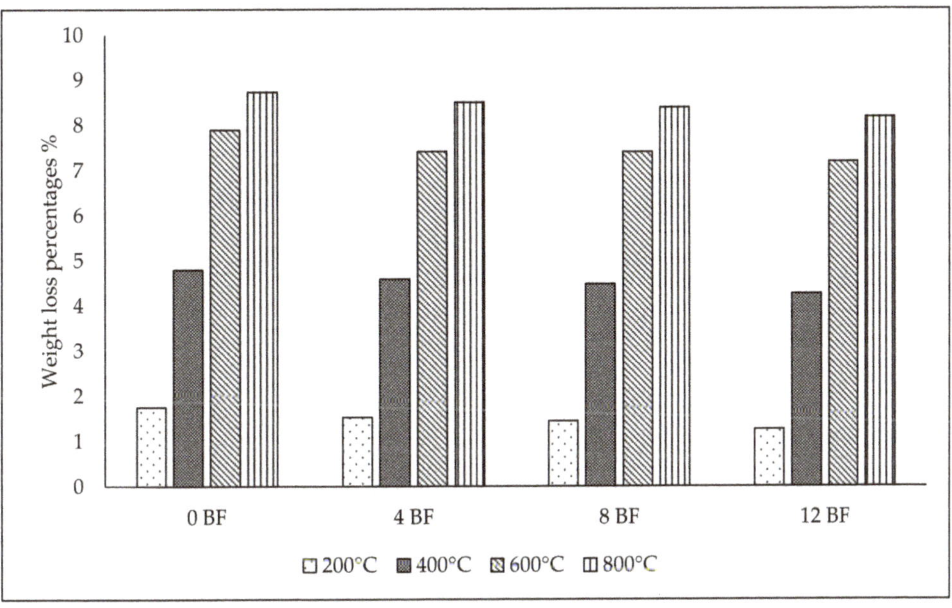

Figure 10. The weight loss percentages of fly ash-based geopolymer samples exposed to high temperatures.

3.5.2. Compressive and Flexural Strength Results

The Compressive strength and flexural strength tests were performed on geopolymer mortar samples which exposed to 200 °C, 400 °C, 600 °C, and 800 °C. Then the results of the samples conducted after and before the high-temperature test were compared. The strength tests were conducted to study the effect of high temperature on the fly ash-based geopolymer mortars' mechanical properties contained BF. Figures 11 and 12 show the results of the compressive and flexural strength of the fly ash-based geopolymer samples. The 12 BF mix had the highest compressive and flexural strength results, and the 0 BF mix had the lowest compressive and flexural strength results. Thus, the BF improved the mechanical properties of the geopolymer samples. The increases of the results of 4 BF, 8 BF, 12 BF samples exposed to 800 °C were 9.01%, 12.79%, and 15.79% for compressive strength and 11.61%, 32.87%, and 44.69% for flexural strength, respectively compared to 0 BF samples. In addition, the increases of the results of 12 BF samples exposed to 200 °C, 400 °C, 600 °C, and 800 °C were 14.31%, 17.18%, 18.65%, and 15.79% for compressive strength and 36.94%, 39.95%, 37.80%, and 44.69% for flexural strength, respectively compared to 0 BF samples.

At 200 °C, the flexural and compressive strengths of the samples did not change significantly (Figures 11 and 12), but beyond 400 °C, the reduction rate of compressive and flexural increased more. Endothermic processes caused these reductions. Dehydration and water vapor of free water occurred concurrently with the processes in the matrix structure of the geopolymer [8,35,36]. In addition, the flexural strength decreases due to the high-temperatures effect were lower than the compressive strength decreases. Where the decreases of the strength of the 12 BF samples exposed to 200 °C, 400 °C, 600 °C, and 800 °C were 7.53% 28.17%, 41.36%, and 54.57% for compressive strength and 2.27%, 14.02%, 33.71%, and 44.32%for flexural strength, respectively compared to the samples not exposed to high-temperature test.

The vapor action emphasized the decrease of compressive and flexural strength as the temperature rose. The geopolymer matrix structure's water transformed into vapor whenever the specimens were exposed to high temperatures. Additionally, increasing temperatures make the Al SiO gel composition crystallized in the geopolymer matrix, causing more microcracks [37]. When the temperature surpassed 100 °C, the internal pressure rose steadily. This scenario produced microcracks in the geopolymer specimens due to crystallization stress. Thermal fractures and thermochemical breakdown in crystal lattices were apparent as the temperature rose. While the sample's strength values decreased, it became more brittle, and fragile cracking appeared. In addition, when the temperature surpassed 100 °C, there was a substantial thermal incompatibility between the paste and the aggregate in geopolymer. Increased temperature also caused microcracks in the aggregate-paste interface transition region, resulting in thermal incompatibilities [38].

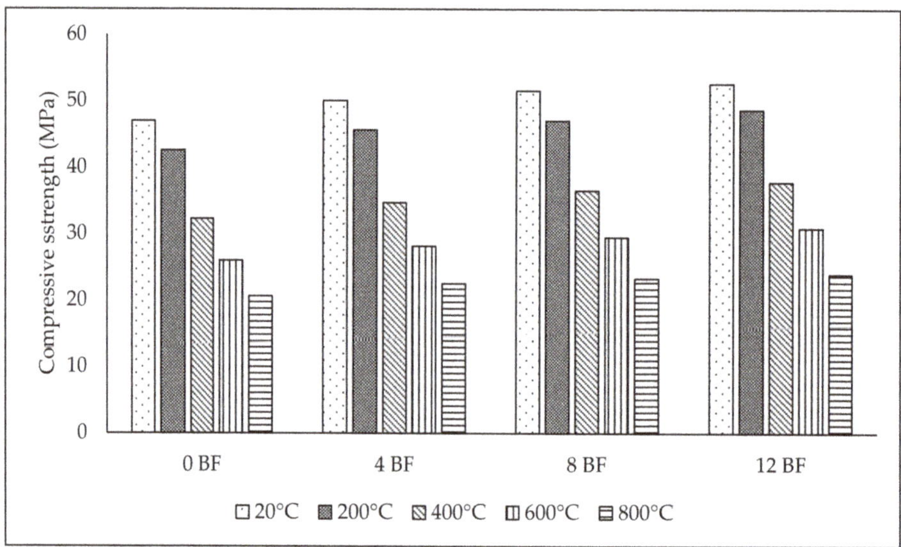

Figure 11. The effect of high temperature on compressive strength of BF contained fly ash-based geopolymer.

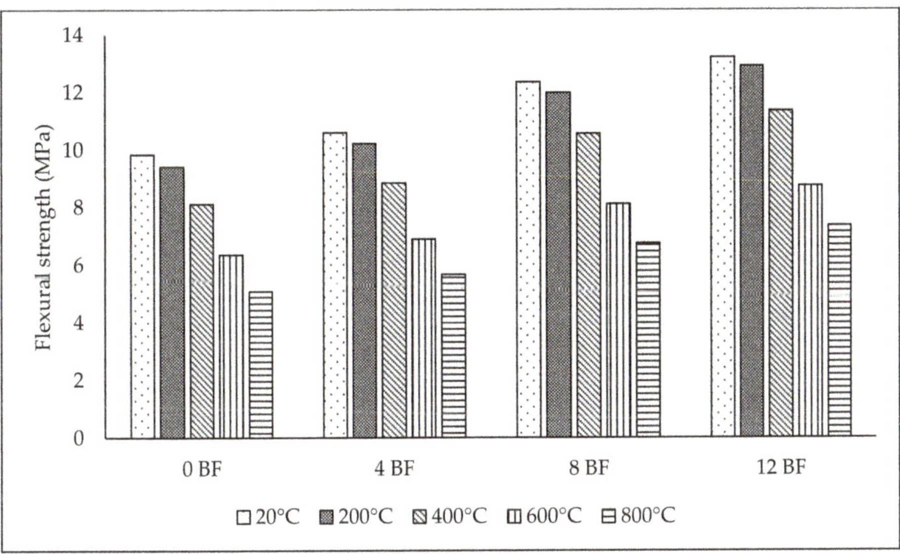

Figure 12. The effect of high temperature on flexural strength of BF contained fly ash-based geopolymer.

3.5.3. Ultrasonic Pulse Velocity (UPV) Results

The UPV test of fly ash geopolymer samples was conducted, and using the UPV results, the damage degrees of samples were calculated (Figure 13). The 12 BF samples had the highest UPV, and the 0 BF had the smallest UPV. Where the UPV results of 0 BF, 4 BF, 8 BF,12 BF samples exposed to 800 °C were 2512.05, 2627.69, 2700.42, and 2785.63 m/s respectively. Thus, the addition of BF in the fly ash geopolymer samples increased the UPV. The UPV readings dropped as the temperature rose for all samples. Due to evaporation in the interior structure, the pore size grew, and the continuity reduced, resulting in this situation [39]. When the UPV findings were compared to the mechanical results, it was observed that the values were almost consistent. At high temperatures, the interface bonds between the aggregates and pastes of the samples weaken, resulting in a reduction in UPV [33]. Where the UPV results of 12 BF samples exposed to 200 °C, 400 °C, 600 °C, and 800 °C were 3312.63, 3190.17, 3020.59, 2905.46, and 2785.63 m/s, respectively.

In addition, the damage degrees of the fly ash-based geopolymer samples exposed to high temperatures were performed. The damage degrees of the fly ash-based geopolymer samples containing BF are shown in Figure 14. The width and number of the cracks of fly ash-based geopolymer samples increased as the high temperatures increased. The UPV values were used to evaluate the degree of damage. In previous studies, Singh and Gupta [40] were used the UPV values to perform the damage degree in their study. A proportional relationship was obtained between the damage degrees and the increase of high temperatures in all series. In addition, as Figure 14 shows, the increase in basalt content decreased the damage degrees of fly ash-based geopolymer mortar samples containing BF with different ratios. The performed damage degrees were ranged between 3.69% for 12 BF samples exposed to 200 °C and 17.42% for 0 BF samples exposed to 800 °C. In other words, the UPV values of the samples exposed to the high-temperature effect were smaller than the initial UPV values of the samples. Thus, high-temperature increases increased the damage degrees of geopolymer samples. The damage degrees of the 12 BF samples at temperatures of 200 °C, 400 °C, 600 °C, and 800 °C were 3.70%, 8.82%, 12.29%, and 15.91%, respectively. This result is compatible with the weight measurement and mechanical strength tests performed in the study. The damage degrees of 0 BF, 4 BF, 8 BF and 12 BF samples exposed to 800 °C were 17.42%, 16.41%, 16.29%, and 15.91%. In addition, the damage degrees of

0 BF, 4 BF, 8 BF, and 12 BF samples exposed to 200 °C were 4.54%, 4.38%, 4.04%, and 3.70%. Thus, the increase in the BF ratio decreased the damage degrees. This situation was because BF inhibited high-temperature cracks in the samples and limited the damage.

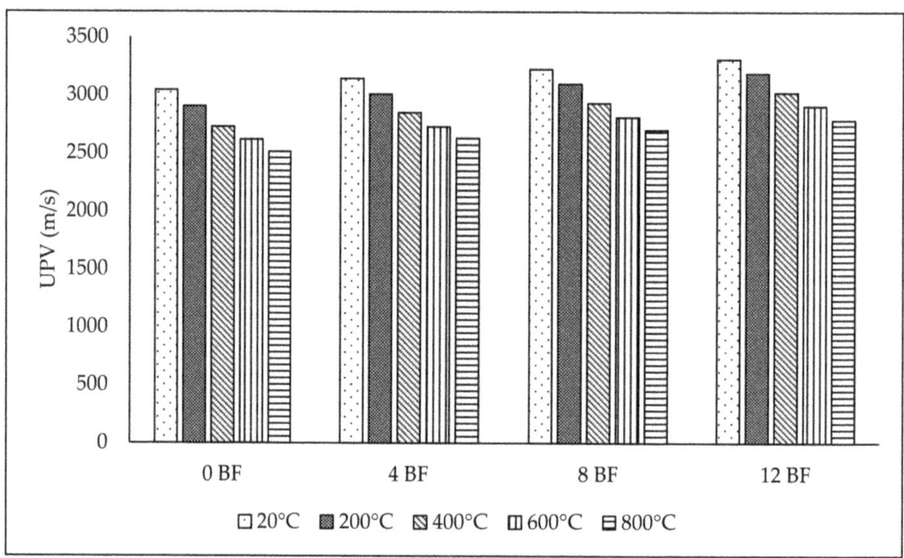

Figure 13. The effect of high temperature on UPV of BF contained fly ash-based geopolymer.

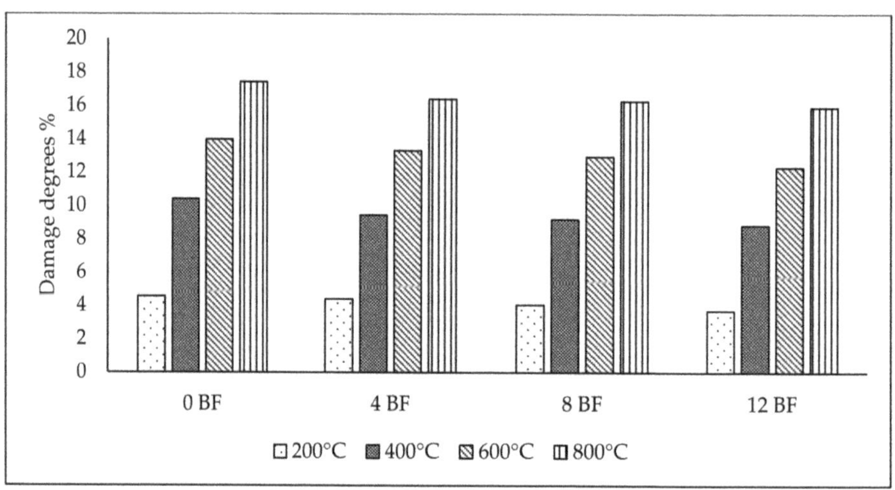

Figure 14. The damage degree of fly ash-based geopolymer samples exposed to high temperature.

3.5.4. Visual Assessment of High-Temperature Test

Fly ash-based geopolymer mortar samples exposed to various temperatures are shown in Figure 15. In general, it was found that the color of the samples and BF exposed to 200 °C and 400 °C did not change compared to the samples not exposed to the high temperature (20 °C). The color of samples exposed to 600 °C and 800 °C changed from gray to brown and white spots were formed in the cross-section. In addition, the BF of the samples exposed to 600 °C and 800 °C changed in color. Unequal volume changes between geopolymer components caused by temperature exposure were referred to as thermal incompatibility of components. This situation also resulted in microcracks in the interfacial transition zone between the geopolymer paste and the aggregate. Microcracks were also increased because of the loss of homogeneity [38].

Figure 15. Fly ash-based geopolymer samples exposed to various temperature effect (**a**) 0 BF; (**b**) 4 BF; (**c**) 8 BF and (**d**) 12 BF.

A 40× magnification microscope was used to examine better the complexity of the outer surfaces of the BF-containing geopolymer samples exposed to high temperatures. The preliminary visual evaluation determined that series 12 BF was the most compact series, as the BF content was higher than the other series. For this reason, 12 BF samples were selected to be examined under the microscope. The microscope images of the 12 BF samples subjected to the high-temperature tests are shown in Figure 16.

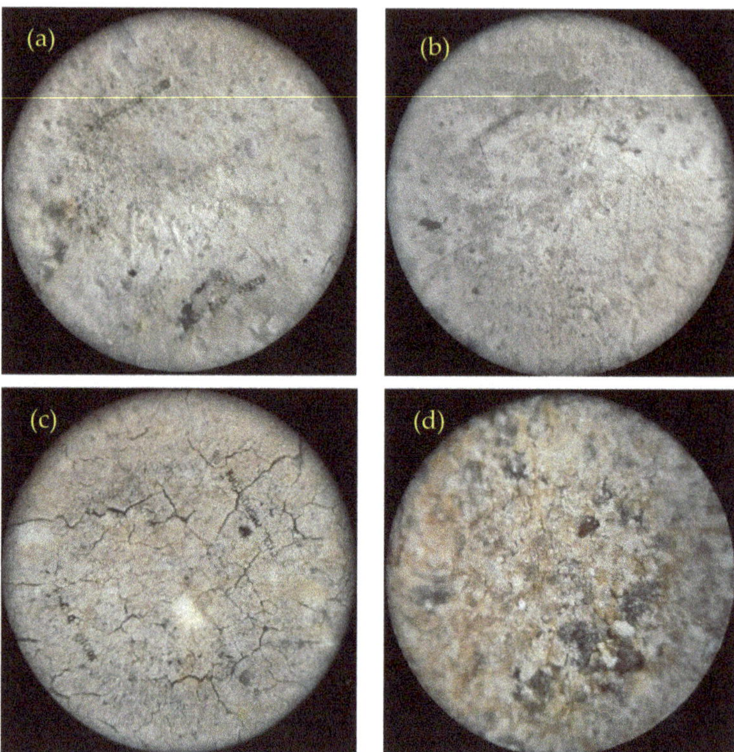

Figure 16. 40x magnified microscope images of 12 BF samples exposed to high-temperature effects (**a**) 200 °C; (**b**) 400 °C; (**c**) 600 °C; (**d**) 800 °C.

Figure 16a,b show that the geopolymer mortar samples exposed to 200 °C and 400 °C did not show any cracks on the surface and the samples maintained their compactness. Figure 16c shows that superficial microcracks formed on the outer surfaces of the geopolymer mortar specimens exposed to 600 °C. Figure 16d shows that the outer surfaces of the specimens exposed to the transition from 600 °C to 800 °C loosened due to superficial cracks, and the cracks were filled with dust formed due to the deteriorated outer surfaces and the aggregates emerged from under the outer surface. As a result, it was observed that the compactness of the geopolymer samples exposed to 600–800 °C deteriorated significantly due to dehydration [8,41].

3.5.5. SEM and EDS Analysis

SEM images of 0 BF and 12 BF samples not exposed to high-temperature tests are shown in Figure 17. As Figure 17a,b shows, 0 BF samples without basalt contained microcracks. The basalt-containing 12 BF samples shown in Figure 17c,d prevented the formation of cracks with the effect of basalt, and it was seen that the 12 BF samples were more compact. The BF acted as a bridging agent, slowing the formation of fractures [42]. In addition, SEM images of both 0 BF and 12 BF samples contain N-A-S-H gel and unreacted FA particles. This finding is consistent with previous fly ash-based geopolymers discovered in the literature [43].

Moreover, SEM images of 12 BF samples exposed to 200 °C, 400 °C, 600 °C, and 800 °C are shown in Figure 18. In Figure 18a,c, it was seen that the structure of the BF of 12 BF samples exposed to 200 °C and 400 °C was the same and did not change. In addition, the 12 BF samples exposed to 200 °C were more compact and had fewer microcracks than the samples exposed to different high temperatures. In this study, the authors were noticed that the matrix of 12 BF samples started to form microcracks after 400 °C and became less compact than the samples exposed to 200 °C. As shown in Figure 18e, voids were formed in the sample exposed to 600 °C due to dehydration and evaporation. Figure 18g shows that the sample exposed to 800 °C had more voids than the sample exposed to 600 °C. Additionally, the voids formed caused volume reduction and structural deterioration of the sample. After exposure to high temperatures, the samples' evaporation rate and water losses increase, leading to structural deterioration and increased formation defects [44].

Moreover, EDS traces of the 0 BF and 12 BF samples not exposed to high-temperature test and 12 BF samples exposed to high-temperature test are shown in Figure 19. Figure 19a,b show that Si/Al ratios of 0 BF and 12 BF fly ash-based geopolymer samples exposed to high-temperature tests are 2.18 and 2.16, respectively. Thus, no significant change was observed in the Si/Al ratio of 12 BF compared to 0 BF. The Si/Al ratio of 12 BF samples exposed to 200 °C, given in Figure 19c, increased to 2.72. Additionally, it was observed that the increase in Si/Al ratio of 12 BF samples exposed to 400 °C given in Figure 19d decreased to 2.41. Figure 19e,f show that the Si/Al ratios of 12 BF samples exposed to 600 °C and 800 °C, respectively, decreased to 2.12 and 2.02. These results are consistent with the study conducted by Abdulkareem et al. [45]. They found that the Si/Al ratio decreased when the fly ash paste samples were exposed to 800 °C.

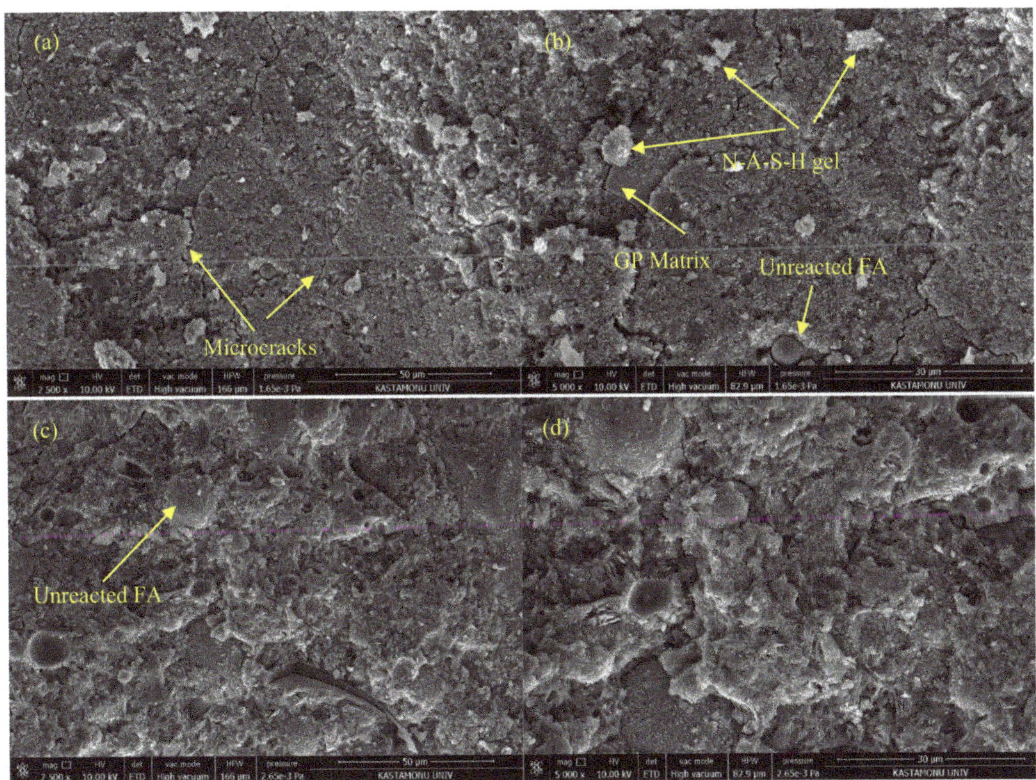

Figure 17. 2500 and 5000 times magnified SEM images of 0 BF sample (**a**,**b**) and 12 BF sample (**c**,**d**).

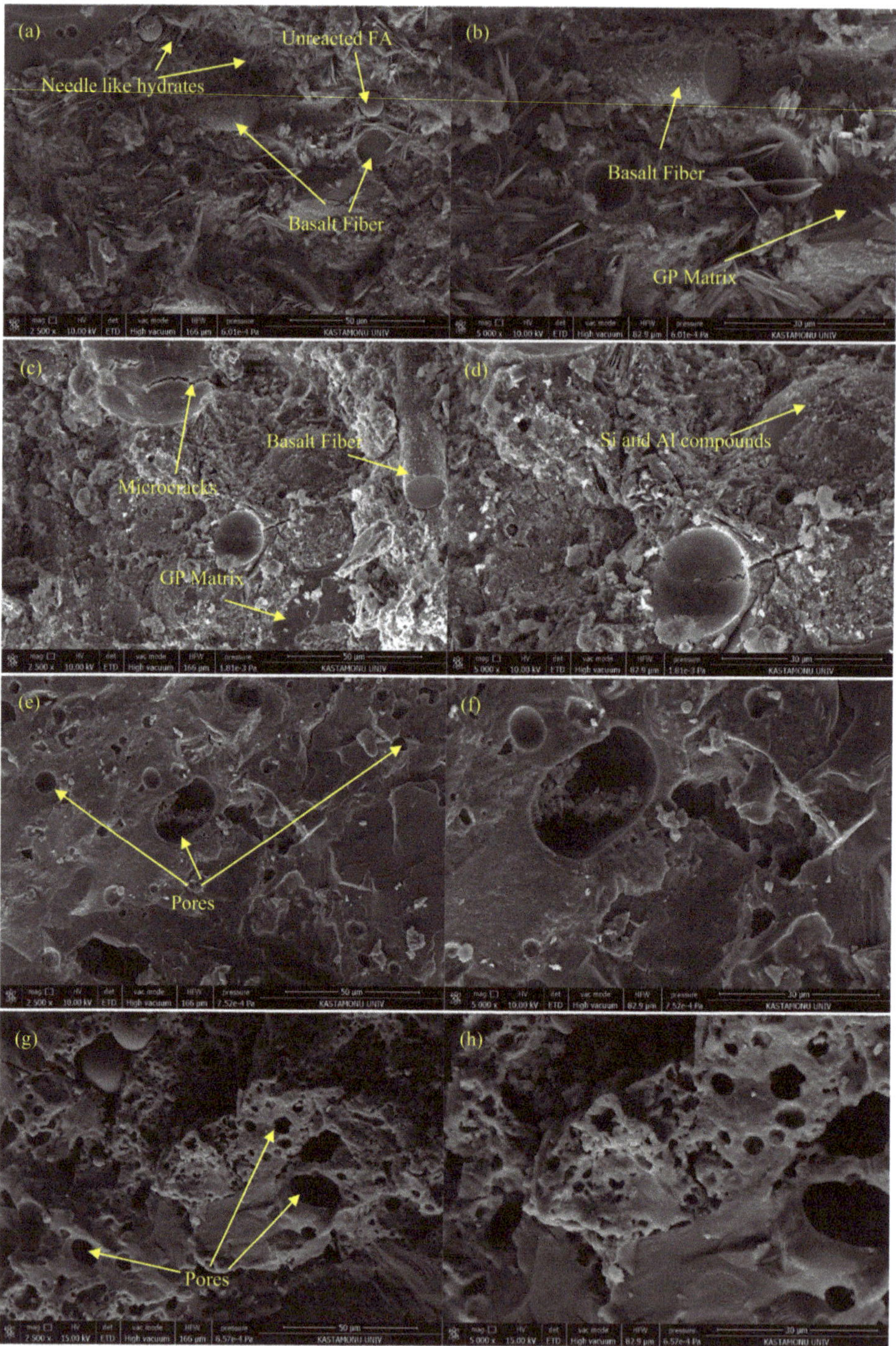

Figure 18. 2500 and 5000 times magnified SEM images of 12 BF sample exposed to 200 °C (**a**,**b**); 400 °C (**c**,**d**); 600 °C (**e**,**f**) and 800 °C (**g**,**h**).

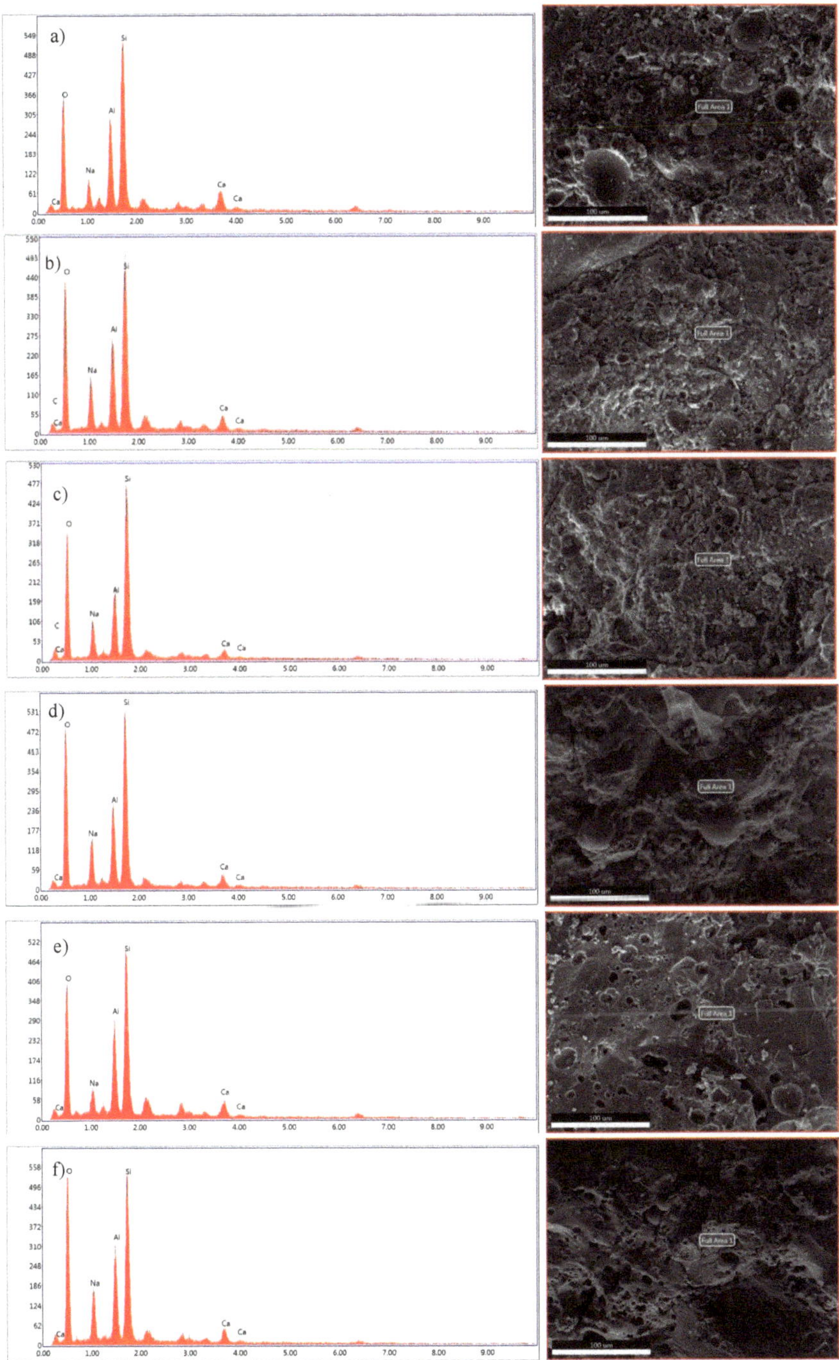

Figure 19. EDS traces of (**a**) 20 °C exposed 0 BF sample; (**b**) 20 °C exposed 12 BF sample; (**c**) 200 °C exposed 12 BF sample; (**d**) 400 °C exposed 12 BF sample; (**e**) 600 °C exposed 12 BF sample and (**f**) 800 °C exposed 12 BF sample.

4. Conclusions

This research examined the impact of BF on a geopolymer based on FA and filled with waste BP. The following summarizes the findings of the tests conducted:

- The study revealed that adding BF to fly ash-based geopolymer mortars improved compressive strength after 28 days. Compressive strength increased by 6.43%, 9.65%, and 11.94% for 4 BF, 8 BF, and 12 BF, respectively.
- BF greatly improved flexural strength. The inclusion of BF decreased the number of fractures during the flexural test by transferring the flexural load. Flexural strength increased by 7.72%, 25.512%, and 34.15% for 4 BF, 8 BF and 12 BF after 28 days. Additionally, in the 12 BF series, flexural strength increased 2.86 times larger than compressive strength. Thus, the addition of basalt increased flexural strength more than compressive strength.
- A good relationship was found between Charpy impact test results and BF content. It improved the Charpy impact test results of the geopolymer specimens after 28 days. As compared to series 0 BF, the Charpy impact test results improved by 78%, 109%, and 118% for series 4 BF, 8 BF, and 12 BF, respectively.
- The increase in basalt content of fly ash-based geopolymer mortars affected negatively by increasing the water absorption ability. This increase occurred because of the inclusion of BF, which improved pore connectivity at high volume fractions. Additionally, the water absorption coefficients of 4 BF, 8 BF, and 12 BF samples increased by 9.04%, 10.55%, and 17.99%, respectively.
- The addition of BF slowed down the weight decreases. The 12 BF samples lost the least weight, while the 0 BF samples lost the most. After 800 °C, the weight loss percentages of 0 BF, 4 BF, 8 BF, 12 BF were 8.73%, 8.49%, 8.38%, and 8.17% respectively. At elevated temperatures, the interface bonds between the paste and aggregate deteriorate, resulting in weight loss. Consequently, the increase of high temperatures increased the weight loss percentages of geopolymer samples.
- The BF improved the compressive and flexural strengths of the samples after and before the high-temperature test. The increases of the compressive and flexural strengths of 4BF, 8BF, 12BF samples exposed to 800 °C were 9.01%, 12.79%, and 15.79% for compressive strength and 11.61%, 32.87%, and 44.69% for flexural strength, respectively compared to 0 BF samples. Thus, the high-temperature effect reduced flexural strength more than compressive strength.
- BF inhibited high-temperature cracks and limited the damages. Thus, the increase in basalt content decreased the damage degrees of fly ash-based geopolymer mortar samples. After the high-temperature test, the damage degrees of samples ranged from 3.69 to 17.42% for 12 BF samples. Additionally, the damage degrees of the samples exposed to high-temperature effects increased proportionally with the increase in high temperatures.
- The color of the samples and BF exposed to 200 °C and 400 °C did not change significantly compared to the samples exposed to 20 °C. Additionally, the samples exposed to 600 °C and 800 °C turned from gray to brown. Dehydration also affected the compactness of the geopolymer samples exposed to 600–800 °C.
- After 400 °C, 12 BF samples developed microcracks and became less compact than the 12 BF samples exposed to 200 °C. In addition, the Si/Al ratio of 12 BF samples exposed to 800 °C decreased compared to 12 BF samples exposed to 20 °C. After exposure to 600 °C and 800 °C, the samples' evaporation rate and water loss rise, resulting in structural deterioration.

Author Contributions: Conceptualization, M.Z., S.E., Y.T., S.K. and R.A.G.L.; methodology, M.Z., S.E., Y.T., S.K.; formal analysis, M.Z., S.E., Y.T., S.K.; investigation, M.Z., S.E., Y.T., S.K.; resources, M.Z., S.E., Y.T., S.K., R.A.G.L.; data curation, M.Z., S.E.; writing—original draft preparation, M.Z., S.E., Y.T., S.K.; writing—review and editing, M.Z., S.E. and R.A.G.L.; visualization, M.Z., S.E., Y.T., S.K., S.E.; supervision, R.A.G.L.; project administration, M.Z., S.E., Y.T., S.K., R.A.G.L.; funding acquisition, R.A.G.L. All authors have read and agreed to the published version of the manuscript.

Funding: The authors wish to thank CEU San Pablo University Foundation for the funds dedicated to the Project Ref. MCP21V12 provided by CEU San Pablo University.

Institutional Review Board Statement: Not applicable.

Informed Consent Statement: Not applicable.

Data Availability Statement: The data presented in this study are available on request from the corresponding author.

Conflicts of Interest: The authors declare no conflict of interest.

References

1. Benhelal, E.; Zahedi, G.; Shamsaei, E.; Bahadori, A. Global strategies and potentials to curb CO_2 emissions in cement industry. *J. Clean. Prod.* **2013**, *51*, 142–161. [CrossRef]
2. Wang, J.; Mu, M.; Liu, Y. Recycled cement. *Constr. Build. Mater.* **2018**, *190*, 1124–1132. [CrossRef]
3. Ma, C.-K.; Awang, A.Z.; Omar, W. Structural and material performance of geopolymer concrete: A review. *Constr. Build. Mater.* **2018**, *186*, 90–102. [CrossRef]
4. Ai, T.; Zhong, D.; Zhang, Y.; Zong, J.; Yan, X.; Niu, Y. The Effect of Red Mud Content on the Compressive Strength of Geopolymers under Different Curing Systems. *Buildings* **2021**, *11*, 298. [CrossRef]
5. Hattaf, R.; Aboulayt, A.; Samdi, A.; Lahlou, N.; Touhami, M.O.; Gomina, M.; Moussa, R. Reusing Geopolymer Waste from Matrices Based on Metakaolin or Fly Ash for the Manufacture of New Binder Geopolymeric Matrices. *Sustainability* **2021**, *13*, 70. [CrossRef]
6. Nikoloutsopoulos, N.; Sotiropoulou, A.; Kakali, G.; Tsivilis, S. Physical and Mechanical Properties of Fly Ash Based Geopolymer Concrete Compared to Conventional Concrete. *Buildings* **2021**, *11*, 178. [CrossRef]
7. Occhicone, A.; Vukčević, M.; Bosković, I.; Ferone, C. Red Mud-Blast Furnace Slag-Based Alkali-Activated Materials. *Sustainability* **2021**, *13*, 1298. [CrossRef]
8. Tammam, Y.; Uysal, M.; Canpolat, O. Effects of alternative ecological fillers on the mechanical, durability, and microstructure of fly ash-based geopolymer mortar. *Eur. J. Environ. Civ. Eng.* **2021**, 1–24. [CrossRef]
9. Çelikten, S.; Atabey, İ. The effects of water content and thermal curing time on physical and mechanical properties of waste basalt powder-based geopolymer mortars. *Ömer Halisdemir Univ. J. Eng. Sci.* **2020**, *10*, 328–332. [CrossRef]
10. Assaedi, H.; Alomayri, T.; Shaikh, F.U.A.; Low, I.-M. Characterisation of mechanical and thermal properties in flax fabric reinforced geopolymer composites. *J. Adv. Ceram.* **2015**, *4*, 272–281. [CrossRef]
11. Shen, D.; Liu, X.; Li, Q.; Sun, L.; Wang, W. Early-age behavior and cracking resistance of high-strength concrete reinforced with Dramix 3D steel fiber. *Constr. Build. Mater.* **2019**, *196*, 307–316. [CrossRef]
12. Vairavan, M.; Pandian, A.; Manikandan, V.; Jappes, T.W.; Uthayakumar, M. Effect of fibre length and fibre content on mechanical properties of short basalt fibre reinforced polymer matrix composites. *Mater. Phys. Mech.* **2013**, *16*, 107–117.
13. Zhang, P.; Gao, Z.; Wang, J.; Wang, K. Numerical modeling of rebar-matrix bond behaviors of nano-SiO_2 and PVA fiber reinforced geopolymer composites. *Ceram. Int.* **2021**, *47*, 11727–11737. [CrossRef]
14. Zhang, P.; Wang, K.; Wang, J.; Guo, J.; Hu, S.; Ling, Y. Mechanical properties and prediction of fracture parameters of geopolymer/alkali-activated mortar modified with PVA fiber and nano-SiO_2. *Ceram. Int.* **2020**, *46*, 20027–20037. [CrossRef]
15. Zhang, P.; Wang, K.; Wang, J.; Guo, J.; Ling, Y. Macroscopic and microscopic analyses on mechanical performance of metakaolin/fly ash based geopolymer mortar. *J. Clean. Prod.* **2021**, *294*, 126193. [CrossRef]
16. Zhang, P.; Zheng, Y.; Wang, K.; Zhang, J. A review on properties of fresh and hardened geopolymer mortar. *Compos. Part B Eng.* **2018**, *152*, 79–95. [CrossRef]
17. Fiore, V.; Scalici, T.; Di Bella, G.; Valenza, A. A review on basalt fibre and its composites. *Compos. Part B Eng.* **2015**, *74*, 74–94. [CrossRef]
18. Girgin, Z.C.; Yıldırım, M.T. Usability of basalt fibres in fibre reinforced cement composites. *Mater. Struct.* **2016**, *49*, 3309–3319. [CrossRef]
19. Du, Q.; Cai, C.; Lv, J.; Wu, J.; Pan, T.; Zhou, J. Experimental Investigation on the Mechanical Properties and Microstructure of Basalt Fiber Reinforced Engineered Cementitious Composite. *Materials* **2020**, *13*, 3796. [CrossRef] [PubMed]
20. Girgin, Z.C. Effect of slag, nano clay and metakaolin on mechanical performance of basalt fibre cementitious composites. *Constr. Build. Mater.* **2018**, *192*, 70–84. [CrossRef]

21. Sadrmomtazi, A.; Tahmouresi, B.; Saradar, A. Effects of silica fume on mechanical strength and microstructure of basalt fiber reinforced cementitious composites (BFRCC). *Constr. Build. Mater.* **2018**, *162*, 321–333. [CrossRef]
22. Wang, Z.; Yu, Q.; Ao, Q.; Du, Q. Mechanical properties of engineered cementitious composite containing basalt fibre. *FEB Fresenius Environ. Bull.* **2020**, *29*, 1997.
23. Zhang, N.; Zhou, J.; Ma, G.-W. Dynamic properties of strain-hardening cementitious composite reinforced with basalt and steel fibers. *Int. J. Concr. Struct. Mater.* **2020**, *14*, 44. [CrossRef]
24. Ali, N.; Canpolat, O.; Aygörmez, Y.; Al-Mashhadani, M.M. Evaluation of the 12–24 mm basalt fibers and boron waste on reinforced metakaolin-based geopolymer. *Constr. Build. Mater.* **2020**, *251*, 118976. [CrossRef]
25. Görhan, G.; Aslaner, R.; Şinik, O. The effect of curing on the properties of metakaolin and fly ash-based geopolymer paste. *Compos. Part B Eng.* **2016**, *97*, 329–335. [CrossRef]
26. Rill, E.; Lowry, D.; Kriven, W. Properties of Basalt Fiber Reinforced Geopolymer Composites. In *Strategic Materials and Computational Design: Ceramic Engineering and Science Proceedings*; The American Ceramic Society: Westerville, OH, USA, 2010; pp. 57–67.
27. Binici, H.; Aksogan, O. Durability of concrete made with natural granular granite, silica sand and powders of waste marble and basalt as fine aggregate. *J. Build. Eng.* **2018**, *19*, 109–121. [CrossRef]
28. Özkan, Ş.; Çoban, Ö. The hybrid effects of basalt and PVA fiber on properties of a cementitious composite: Physical properties and non-destructive tests. *Constr. Build. Mater.* **2021**, *312*, 125292. [CrossRef]
29. Kim, D.J.; Wille, K.; El-Tawil, S.; Naaman, A.E. Testing of cementitious materials under high-strain-rate tensile loading using elastic strain energy. *J. Eng. Mech.* **2011**, *137*, 268–275. [CrossRef]
30. Ma, Y.; Zhu, B.; Tan, M. Properties of ceramic fiber reinforced cement composites. *Cem. Concr. Res.* **2005**, *35*, 296–300. [CrossRef]
31. Wang, Y.; Hughes, P.; Niu, H.; Fan, Y. A new method to improve the properties of recycled aggregate concrete: Composite addition of basalt fiber and nano-silica. *J. Clean. Prod.* **2019**, *236*, 117602. [CrossRef]
32. Şahin, F.; Uysal, M.; Canpolat, O.; Aygörmez, Y.; Cosgun, T.; Dehghanpour, H. Effect of basalt fiber on metakaolin-based geopolymer mortars containing rilem, basalt and recycled waste concrete aggregates. *Constr. Build. Mater.* **2021**, *301*, 124113. [CrossRef]
33. He, P.; Jia, D.; Lin, T.; Wang, M.; Zhou, Y. Effects of high-temperature heat treatment on the mechanical properties of unidirectional carbon fiber reinforced geopolymer composites. *Ceram. Int.* **2010**, *36*, 1447–1453. [CrossRef]
34. Kong, D.L.Y.; Sanjayan, J.G.; Sagoe-Crentsil, K. Comparative performance of geopolymers made with metakaolin and fly ash after exposure to elevated temperatures. *Cem. Concr. Res.* **2007**, *37*, 1583–1589. [CrossRef]
35. Arslan, A.A.; Uysal, M.; Yılmaz, A.; Al-mashhadani, M.M.; Canpolat, O.; Şahin, F.; Aygörmez, Y. Influence of wetting-drying curing system on the performance of fiber reinforced metakaolin-based geopolymer composites. *Constr. Build. Mater.* **2019**, *225*, 909–926. [CrossRef]
36. Aygörmez, Y.; Canpolat, O.; Al-mashhadani, M.M.; Uysal, M. Elevated temperature, freezing-thawing and wetting-drying effects on polypropylene fiber reinforced metakaolin based geopolymer composites. *Constr. Build. Mater.* **2020**, *235*, 117502. [CrossRef]
37. Zhang, H.Y.; Kodur, V.; Wu, B.; Cao, L.; Qi, S.L. Comparative thermal and mechanical performance of geopolymers derived from metakaolin and fly ash. *J. Mater. Civ. Eng.* **2016**, *28*, 04015092. [CrossRef]
38. Jiang, X.; Xiao, R.; Zhang, M.; Hu, W.; Bai, Y.; Huang, B. A laboratory investigation of steel to fly ash-based geopolymer paste bonding behavior after exposure to elevated temperatures. *Constr. Build. Mater.* **2020**, *254*, 119267. [CrossRef]
39. Topçu, İ.B.; Karakurt, C. Properties of reinforced concrete steel rebars exposed to high temperatures. *Res. Lett. Mater. Sci.* **2008**, *2008*, 814137. [CrossRef]
40. Singh, H.; Gupta, R. Cellulose fiber as bacteria-carrier in mortar: Self-healing quantification using UPV. *J. Build. Eng.* **2019**, *28*, 101090. [CrossRef]
41. Celik, A.; Yilmaz, K.; Canpolat, O.; Al-mashhadani, M.M.; Aygörmez, Y.; Uysal, M. High-temperature behavior and mechanical characteristics of boron waste additive metakaolin based geopolymer composites reinforced with synthetic fibers. *Constr. Build. Mater.* **2018**, *187*, 1190–1203. [CrossRef]
42. Xu, J.; Kang, A.; Wu, Z.; Xiao, P.; Gong, Y. Effect of high-calcium basalt fiber on the workability, mechanical properties and microstructure of slag-fly ash geopolymer grouting material. *Constr. Build. Mater.* **2021**, *302*, 124089. [CrossRef]
43. Temuujin, J.; Minjigmaa, A.; Rickard, W.; Lee, M.; Williams, I.; Van Riessen, A. Fly ash based geopolymer thin coatings on metal substrates and its thermal evaluation. *J. Hazard. Mater.* **2010**, *180*, 748–752. [CrossRef] [PubMed]
44. Zhang, Y.; Li, S.; Wang, Y.; Xu, D. Microstructural and strength evolutions of geopolymer composite reinforced by resin exposed to elevated temperature. *J. Non-Cryst. Solids* **2012**, *358*, 620–624. [CrossRef]
45. Abdulkareem, O.A.; Mustafa Al Bakri, A.M.; Kamarudin, H.; Khairul Nizar, I.; Saif, A.A. Effects of elevated temperatures on the thermal behavior and mechanical performance of fly ash geopolymer paste, mortar and lightweight concrete. *Constr. Build. Mater.* **2014**, *50*, 377–387. [CrossRef]

Article

Sustainable Construction: Improving Productivity through Lean Construction

Tamar Awad *, Jesús Guardiola and David Fraíz *

Department of Mechanical Engineering, Technical School of Engineering at ICAI, Comillas University, 28015 Madrid, Spain; jguardiola@comillas.edu
* Correspondence: tawad@comillas.edu (T.A.); dfraiz@comillas.edu (D.F.)

Abstract: The objective of this article is to improve building productivity, evolving from traditional construction to industrial construction. The methodology used here consists of analysing the use of materials, the construction design, the design of the spatial distribution programme, the use of auxiliary means and resources and the application of lean tools in construction. The results achieved here include a complete building system that integrates the design, project and execution, wherein the construction process is improved and inconsistencies in the final project are reduced. With the application of an industrial manufacturing methodology, the productivity in construction is improved, reducing costs, materials, execution times and waste. These productivity improvements result in construction being more sustainable. As a conclusion of the previous analysis, the elements that must be integrated into a complete building project and the need to incorporate industrial manufacturing into construction processes in order to achieve sustainable architecture are established.

Keywords: lean manufacturing; modular construction; sustainability architecture; efficient buildings; sustainability; lean construction

Citation: Awad, T.; Guardiola, J.; Fraíz, D. Sustainable Construction: Improving Productivity through Lean Construction. *Sustainability* **2021**, *13*, 13877. https://doi.org/10.3390/su132413877

Academic Editors: Roberto Alonso González Lezcano and Marc A. Rosen

Received: 26 October 2021
Accepted: 8 December 2021
Published: 15 December 2021

Publisher's Note: MDPI stays neutral with regard to jurisdictional claims in published maps and institutional affiliations.

1. Introduction

The intention of this article is to study construction in order to propose a sustainable method in which it can be carried out. One possible way to achieve this is through the application of lean manufacturing in construction processes.

The lean concept [1] was born in the 1930s through the observation of processes in Toyota factories [2] and, since then, it has typically been applied in productive or industrial processes through the use of different techniques and tools, such as 5s, SMEN, QFD, TPM, Kamban and Pocayoke, etc., achieving adjusted production. With sustained applications over time, the continuous improvement and optimisation of different productive processes are achieved.

For this process, the lean methodology seeks the reduction of "waste," commonly in seven categories: Overproduction, waiting time, transport, excess procedures, inventory, movements, defects and the non-contribution of the operator in obtaining ideas that can improve processes.

As can be seen, this waste produced in the manufacturing industry is easily extrapolated to the building sector, as the construction process is considered to be an industrial or productive process [3].

Combining the use of the lean methodology in construction [4] and BIM (Building Information Modelling) [5] for integrated building management would be the best way to improve sustainability in construction, thereby achieving certain SDGs goals, such as 9, 11 and 12 [6].

The production process in construction should be understood as an integrated process as are other industrial activities. However, at present, it is very fragmented, categorised as independent fields such as design, focused on the speciality of architecture; the execution

141

of the project, split between architects and engineers; and construction, carried out by independent companies separate from the fields of action of those previous, where architects and engineers are also present.

When considering any other process and development in the industrial field, the activities, tasks, and work are coordinated, managed, and directed in a very different way to building processes.

Construction, in order to be included in an industrial dynamic, should combine the following stages: Design, devising the space, the project, the transfer to production, and finally, the construction of the work, as well as the post-sale follow-up.

An analysis of today's buildings has many characteristic aspects that are different from typical industrial production, where all production activities are usually controlled by the same entity or company and fall under the same methodology, standards, regulations, and common performance objectives.

Lean techniques can be applied to all steps of the construction process, which encompass lean project definition, lean design, lean supply, lean assembly, and others [7]. The focus is on how a lean design and lean assembly can be combined to contribute to a sustainable production process.

The production of buildings is carried out in three phases: Design, engineering, and assembly [8].

In development, there are the following three stages:

- Design: Focusing on the treatment of the habitat and functional conditions of the building.
- Engineering: Referring to the preparation of the design for construction.
- Assembly: On-site construction of the building.

The results of this development are oriented towards obvious economic advantages, such as a cost reduction, which is socially necessary. The deadline, which is an aspect directly linked to the cost, will also be addressed more than the economic aspect.

Other advantages of the reorientation of building to an industrial environment include quality assurance, sustainability [9] and digitalisation.

Unlike manufacturing, construction is a project-based production process. Lean construction [10] is concerned with the integral pursuit of simultaneous and continuous improvements in all dimensions of the built and natural environment: Design, construction, commissioning, maintenance, recovery, and recycling.

2. Materials and Methods

The methodology is based on an analysis of the elements involved in the design and construction process. Through the use of lean tools, productivity is improved, which results in construction being sustainable. The following five aspects are developed below: An analysis of the current situation, the discourse of the method, an analysis of UNE-EN 15643-3-2012, the definition of the programme and the meaning of construction as engineering through a price decomposition analysis.

Firstly, an analysis is conducted of the current state of the art [2], the traditional building processes and the challenges of new buildings related to the industrialisation available. We also analysed the advances achieved and the possible solutions for the application in the construction industry.

2.1. Analysis of the Current Situation: Industrial versus Traditional Construction

Since the end of the 1980s, the lean philosophy has been adopted in the construction sector with a focus on efficiency, mainly motivated by economic competition [11].

In terms of industrialisation, there are differences between products and technology. Products refer to construction components available on the market, and technologies to the methodologies for on-site assembly.

2.1.1. Products and Technologies

Nowadays, sufficient materials, products and systems are available to develop a more advanced building technology [12] than traditional building systems of the past.
Products:

- In situ: Made on-site, semi-finished, they require time to dry or set, they also need auxiliary means, formwork, falsework . . .
- Workshop: custom-built off-site, available unassembled and ready to install on-site.
- Factory: Standardised, ready to be assembled on-site, adjusts to measurements easily.

Technologies:

- Craftsmanship: Labour-intensive on-site assembly, bonding with binders and adhesives.
- Industrial: Labour-simplified assembly, with standards or instructions, dry assembly by screwing, adjusting, skilled and specialised labour.

Figure 1 is a study of the combination possibilities between the origin of materials and the technology used.

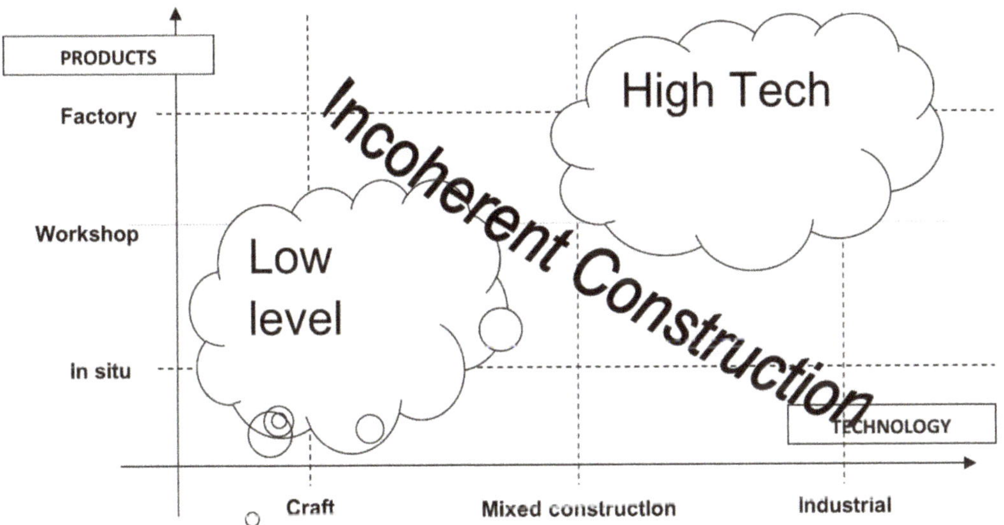

Figure 1. Products and technology (own source).

Incoherent construction happens when technology and product categories that do not correspond are mixed. Outside the "cloud" areas, advanced industrial design technological products are combined with traditional, archaic construction.

2.1.2. Technology and Design

This is an analysis of building systems and technologies and their connection to the elements of programme design and space distribution.

Sometimes, architecture that is intended to use advanced construction systems is paradoxically executed with handcrafted construction. This results in incoherence between design and construction.

Two construction systems can be distinguished as shown in Figure 2.

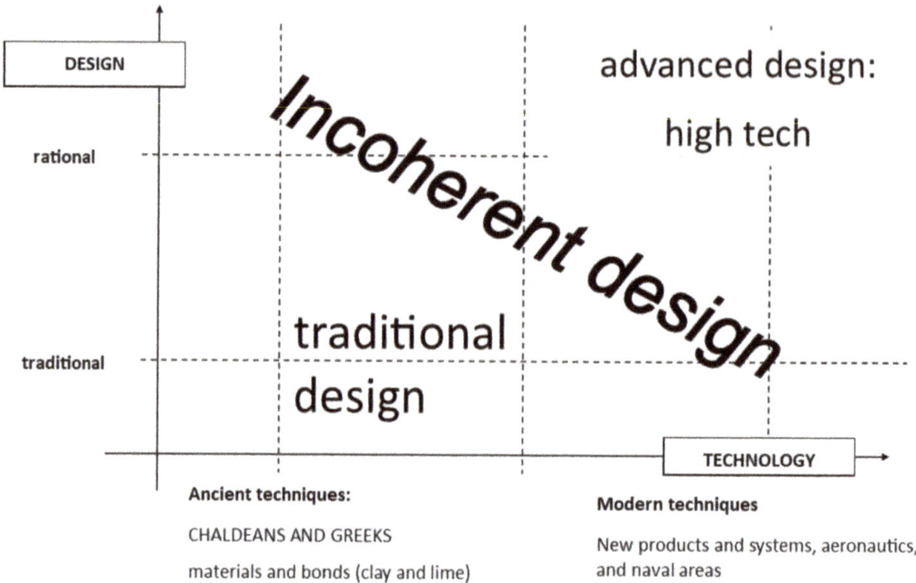

Figure 2. Design and technology (own source).

System based on ancient techniques and materials: from archaic eras, based on earthy materials and hydraulic joining methods, which combine water with binders that set over time.

Modern systems based on external technologies: Aeronautical, naval [13] and others, which provide lighter materials, composites of elements with combined properties, according to more optimized requirements, as well as more rationalised execution methods.

There are four possible outcomes distinguished by the design of the habitat and the construction technology provided.

- Rational and contemporary design combined with advanced technology: High tech [14].
- Conservative and outdated design with archaic construction: Traditional design.
- Rational design using traditional construction technology.
- Conservative design using advanced technology.

The first two possibilities are coherent and parallel between the design criteria and the technology chosen for their execution; however, the latter two produce an incoherent result.

2.2. Stages of the Construction Project

Once the current situation has been analysed, reviewing the building project [6] can lead to an improvement in the construction process, resulting in a better level of quality, lower cost, and shorter execution time.

The complete cycle is reviewed—design, project, and assembly—in order to obtain a better performance in terms of cost and construction time. This process review, which is carried out within the lean construction methodology [15], seeks the optimisation of each process of components, resources and materials.

According to Koskela, there are two main processes in a construction project:

- Design process: this is a step-by-step refinement of specifications where vague needs and wishes are transformed into requirements, and after a variable number of steps, into detailed designs.
- Construction process: Composed of two different types of flows; material process and construction equipment work processes.

Processes can be characterised by their cost, duration, and value to the customer [16].

A lean analysis in construction must be based on the order of stages established in construction projects, as set out in the UNE-EN 15643-3-2012 standard: Terminology of Engineering Services in Buildings, Infrastructures, and Industrial Installations.

To analyse the complete construction process, from the design to the material execution, the description of the UNE-EN 15643-3-2012 standard is used.

UNE-EN 15643-3-2012 is a reference to sustainability in construction.

From the following chart, Figure 3, we have chosen the following stages of development of a construction project. Those that will best fit the above-mentioned arguments of "design" and "material execution" are the following:

0. Initiative:

    0.1. Market Study
    0.2. Business Case

1. Start:

    1.1. Project Start
    1.2. Viability Study
    1.3. Project Definition

2. Design:

    2.1. Conceptual Design
    2.2. Preliminary Design
    2.3. Technical Design
    2.4. Detail Design

Among the four design stages listed in the UNE-EN 15643-3-2012, the conceptual design and the preliminary design belong to the architecture project group, while technical design and detail design would be in the engineering execution group.

On an architectural level, in the project development phase, we will analyse the architectural design manners of the habitat in order to find lines of improvement in the organisation of the programme.

For the second group—engineering—oriented to the material execution of the project, the economic configuration is studied, with price base and deadlines.

The main variables to be considered in costs are:

- Materials
- Manpower: M.P.
- Auxiliary resources: A.R.

Materials are the products that enter the construction site for assembly or joining. Manpower is the human resources necessary to carry out the work. The auxiliary resources are the means to facilitate such work, tools, and machines.

By analysing the architectural elements, the way to optimise cost efficiency is studied, from the revised design aimed at improving the productivity of execution to the cost components and construction times to be reduced, depending on the construction process itself.

The next subsection is an analysis of each of these two groups, framed in the "architecture" and "engineering" of the construction processes, including the stages set out for each of them, the particular developments of the project, seeking options for cuts and elimination of superfluous elements, typical of lean methodology in construction.

2.3. Design: Definition of the Programme

The industrial engineer [17], who designs cars, railways, aeroplanes, and ships, surpasses the construction engineer by using mechanically processed materials and methods for rational production. The problem has to be approached from three interdependent factors: economy, technology and method.

EN 15643-3-2012		STAGES	SUBSTAGES		CUSTOMER APPROVAL BY SUB-STAGES based on assessments of cost, organisation, scheduling, information, quality, risk, environmental impact, viability, etc.	Application for additional authorisation
BEFORE THE USE STAGE	PRODUCT STAGE	0. INITIATIVE	0.1.	Market study		
			0.2.	Business case		
		1. START	1.1.	Project start		
			1.2.	Viability study		
			1.3.	Project definition		
		2. DESIGN	2.1.	Conceptual design		
			2.2.	Preliminary design and developed design		
			2.3.	Technical design		
			2.4.	Detail design		
		3. PROCUREMENT	3.1.	Procurement		
			3.2.	Construction contract		
	CONSTRUCTION STAGE	4. CONSTRUCTION	4.1.	Pre-construction		
			4.2.	Construction		
			4.3.	Implementation		
			4.4.	Delivery		
			4.5.	Statutory approval		
STAGE OF USE		5. USE	5.1.	Operation		
			5.2.	Maintenance		
END-OF-LIFE STAGE		6. END OF LIFE	6.1.	Refurbishment		
			6.2.	Dismantling		

Figure 3. Based on UNE-EN 15643-3-2012: sustainability in construction: sustainability assessment in buildings.

The economic factor consists of analysing cost reduction, the technological one of providing the materials, products, and components, all based on mass production means, and the method, which deals with the rational composition of the living space and its design, on the other hand.

From the historical analysis of habitat needs, two common strategies can be observed: the revision of the way of living, its needs or corresponding consumption of space and the search for technologies that facilitate its execution, with the aim of economising costs and economic effort.

Thus, from new solutions and revised approaches to the habitat scheme, the aim is to achieve a new rational industrialised building.

Before moving on to industrial production in construction, the requirements and use of the spaces have to be clear.

The design of space is looking for housing with rational use. As a result of this, there are habits that have to be revised, rationalising the organisation of living space.

A building system that combines design and execution will facilitate the typical construction processes, based on industrial construction engineering, transferring methodologies that improve the productivity of construction on-site, and that will optimize lean construction.

This section begins to focus the analysis on the design of the "product" itself, which is the habitat.

2.3.1. Functional Analysis

Facing a new form of construction affects both the specific execution process and the design of the space.

It is necessary to delve into the design from the beginning, analysing the habitat or way of living, its relations and internal functions and its needs and uses; in short, its functional programme.

The in-depth study of this section is enormously complex as it is an extensive field of analysis, in all its cultural, sociological, and psychological components; many authors, analysts and architecture professionals have been working for some time on the exclusive subject of housing, including its political implications since the last century, with enormously committed positions.

In this sense, the CIAM congresses, Le Corbusier's Athens Charter [18] and all that follow are both in favour and broad critical positions: Team X and Hadraken's diagrammatic methodology [19], Japanese Metabolists [20], and Archigram [21]. Up to the present day, this is a debate that is still unfinished.

The issue of housing has long been a problem of difficult satisfaction for society.

At the beginning of the 20th century, the social and economic panorama began to change, with the studies of the Industrial City by Tony Garnier [22], in which new approaches to an integrated habitat were presented.

From the early years of Russian constructivism, through the successive trends of modern architecture, solutions of housing units have been collected (Le Corbusier), in the approaches of minimal housing, on arguments of hygiene and health or in the revision of a new way of life, open to nature and today's society, for example, Los Angeles architecture with case study houses.

Likewise, even designs have been revised to facilitate a better and more rational construction: the five points of Le Corbusier, Bauhaus school [23], Prouvé, even in Spain, the contribution of Rafael de la Hoz [24] stands out, looking for solutions in architectural elements.

On the other hand, pioneering engineers, using new materials such as steel, concrete, plastics, and laminated wood, in line with the industrialisation of building, such as E.L. Ransome [25,26] in the use of prefabricated concrete parts in the USA or Freyssinet in France, with the evolution of pre-stressed concrete [27], provide applications in civil and industrial fields.

Governments, city councils and schools of design and architecture universities have always been gradually getting involved, looking for the solution to the population's need for housing. This began with the pressing problems that arose after the Second World War, where Europe lost an enormous number of buildings and housing and looked for quick and emergency solutions, from which the first applications of mass prefabrication [28] emerged but did not turn out to be the desired solution.

2.3.2. Rational Design: The Architectural Project

This section analyses the minimum elements necessary for the habitat.

There is no justification for the fact that every house in a suburban neighbourhood has a different typology: different floor plans, facades and building materials. Throughout Europe, the old farmhouse, and the townhouse of the average citizen in the 18th century show a similar layout.

Except for the typologies revised during the modern movement, in search of the minimum house solution, the way of living is still traditional, falsely conditioned by the construction of load-bearing walls, where rooms are enclosed between walls and elongated corridors.

Housing should be designed according to the requirements of the family unit but tending to standardise the common parts, eliminating unnecessary uses and spaces and generalising related spaces.

The optimisation of the habitat focuses on the strategy of reducing architectural components through function and use, using architectural elements that configure spaces according to their use.

According to A. Moles, the necessary minimum elements of a space configuration can be determined as [29]:

A.   The sphere of the "immediate gesture": It is made up of the furniture close to each function.

B.   The "room," the minimum determined space, configured by each function in a space, delimited by partitions.

C.   The "flat," the composition of complete functions of the habitat, configuring the hierarchy of complete functions of a habitat.

In this strategy of minimums can be referenced the design modes for a revised habitat result, with the saving of architectural resources, which ultimately saves costs, preserving user value.

As an example, we can propose solutions, taken from the courses at the Madrid School of Architecture, taught by A. Fernández Alba in the 1970s [30].

In these solutions, the concepts of minimal conceptual strategies applied to the proposed spaces are tested (Figure 4).

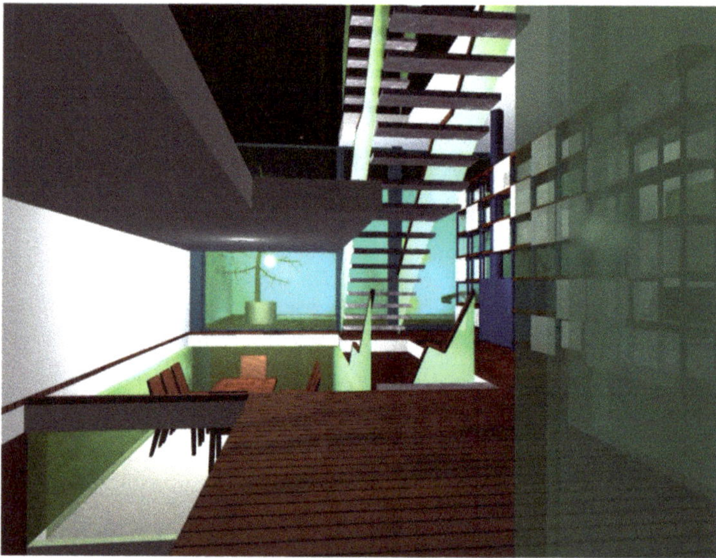

Figure 4. Interior perspective: rehearsal of the organisation of living space: J. Guardiola.

Elements of Design

A more controlled design of the habitat conditions, programme, space and uses favours the rationalisation of the execution on the elements of the construction: Walls, installations or façades.

The basic functions can be pointed out, which will help rethink less traditional housing solutions and involve more current approaches among the needs of the habitat.

Functionality

The design premises, at afunctional level, which present the problem of building organisation are set out under the following principles:

- Privacy and community: the private and the common
- Functional spaces

- Sociological changes

Elements Reviewed

A revision of the programme allows a reduction of architectural components, which favours cost-cutting.

Identification of components that make up the functional and technical categories:

- External cladding or skins: façade, roofing
- Supporting plans: floor slabs
- Interior partitions: walls
- Fluid lines: pipelines (electrical, mechanical, gas, etc.)

A rational study of the design of the habitat facilitates its execution, towards economically optimised results, as many superfluous elements are eliminated.

Some components of functional space can be reduced: corridors and partitions for semi-private spaces, while compatible spaces can be unified.

On the other hand, the free plan, as Le Corbusier's paradigm, achieves the simplicity of relations between uses of the building, not conditioned by the rigid structural functions.

A well-coordinated structural choice can simplify internal partitioning conditions and eliminate redundant elements. (Figure 5).

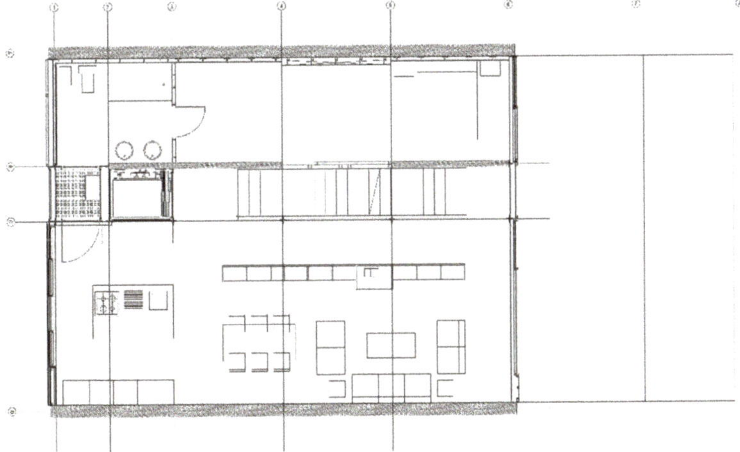

Figure 5. Housing plan: rehearsal of the organisation of a living space–J. Guardiola.

The application of the lean methodology in construction allows for the elimination of components that do not add value to the user, which is achieved by reviewing the needs of the habitat programme in the stages from the beginning of the project [31].

2.4. Construction Engineering: Price Decomposition Analysis

The analysis of productivity improvement in construction focuses on cost elements, not including other aspects such as management, organisation or quality.

The aim of systems engineering [32] is to formalise new ways of approaching the execution of the building, with which we will understand better the advantages of industrial approaches, using its own internal logic in the design.

However, it is not only necessary to industrialise the components for their industrial manufacture but also to go further, thinking about assembly and maintenance, summarized by contemplating the full 360° cycle of the complete exploitation of the building asset.

Cost optimisation is studied based on the price decomposition of the construction components, following the elementary cost analytics known in construction.

As far as execution concerns, including material and assembly, the material execution cost is established, which is broken down into three elementary prices:

- Price of materials.
- Manpower price.
- Price of auxiliary resources.

When analysing the construction timings for the execution process, different types of construction timings are found:

- Coordination times: one task depends on another to continue. Analysing the assembly, looking for solutions of "parallel" tasks and prioritising over "serial" tasks.
- Construction times of semi-finished components that are not finished for their function; looking for "finished" materials that do not require construction times.
- Weather: atmospheric conditions that condition construction (setting, drying); looking for processes and materials, independent of weather conditions.

The deadline is reduced, with the shortening of work times, in each of its modalities.

Therefore, the cost—through its impact of manpower and auxiliary resources—the deadline and the design will be the scope dimensions of this study.

The Elements of the Material Execution Cost: Current Situation

Current construction uses old methods and materials, and the execution is highly conditioned by manpower and weather conditions and is very dependent on auxiliary resources.

The price of a construction element consists of: manpower, materials and auxiliary resources.

- Currently, in construction, the cost of the construction elements is involved in a differentiated manner, as explained below with the impact on the implemented component.
- Auxiliary resources: these are correlated by their time of use, which means that the shorter the time, the lower the cost.
- Materials: traditional, obsolete, archaic technology.

On the other hand, execution times are aggravated by current on-site construction methods, affected by weathering and by a linear sequence of jobs, which accumulate their times.

More extensively, these observations are further elaborated by differentiating for the three cost elements:

- Manpower considerations: Lack of homogeneity. No progress on the technical solution. Construction work is affected by the weather. Improvisation, due to lack of detail in the design. Quantity over quality.
- Auxiliary resources represent two types of units:

Means charged directly on items, in which repercussion, according to companies, is in the range from 1 to 2% and refer to tools applied for internal finishing works, not applicable for installation works.

Special external means for lifting and external platforms and mixing and storing hydraulic materials, through which repercussion is a direct price and varies depending on the construction. These means are normally applied as a temporary rate, either as a rental or as amortisation, and are charged for the time employed, so that the term acts directly against these units.

- Material considerations:

Material adjustments to the geometry are necessary, resulting in waste: shrinkage and trimming.

Traditional materials without technological evolution, natural materials with few possibilities for development.

End caps, overlaps: special pieces to cover up or cover unavoidable defects.

Auxiliary resources required for their work: Moulds, formwork, scaffolding, platforms and cranes, etc.

They require drying and setting times due to the use of hydraulic materials, usually binders and finishing materials.

- Timeframe considerations:

Semi-finished products, which require time for shaping, setting or drying.
Semi-finished products, which require a large amount of manpower to be produced on-site and therefore time consumed.
It is of interest to shift the manpower time from on-site to off-site production: workshop, factory or industrial factory.
Linear construction: chained times in traditional construction, with waiting times between them due to the difficulty of coordination between different trades.

3. Results

In summary, the following results can be deduced from the optimisation of the lean methodology in construction.

Firstly, in the same methodological order, through the results obtained from the analysis of the combination of technologies and products used in construction (Section 2.1.1), it can be determined that incoherent construction occurs when mixing craft technologies with factory-produced materials or industrial technologies using materials made on site.

According to the analysis of technologies and design (Section 2.1.2), there are four possibilities, distinguished between the design of the habitat and the construction technology provided.

- Rational and contemporary design combined with advanced technology: High tech.
- Conservative and outdated design and archaic construction: Traditional design.
- Rational design using a traditional construction technology.
- Conservative design using advanced technology.
- The first two possibilities are coherent and parallel between the design criteria and the technology chosen for their execution; however, the latter two produce an incoherent result.

From the stages of the project, studied in the methodology (Section 2.2), there are two areas in a construction project: The design and the execution.

Of the four design stages listed in the UNE-EN 15643-3-2012, conceptual design and preliminary design would be in the architecture project group, while technical design and detail design would be in the engineering execution group.

We summarize the review of the design definition of the programme (Section 2.3) followed by the results of the economic analysis of the price in construction. It is necessary to reduce architectural elements and resources, materials, and execution time.

3.1. Reduction of Architectural Elements

The results are obtained from the analysis of the construction systems and technologies used in relation to the design.

A rationalised design of the habitats' conditions, programme, space and uses, using arguments of the original design and fundamentally detached from traditional and conservative typologies can lead to innovative solutions along the lines of the savings sought in lean construction.

The rationalisation of construction is favoured on the defining elements of the architectural space, according to three differentiated modes:

- Elimination of corridors, lobbies, and room separations, resulting in consequent savings:
    - Reduction of interior partitions
    - Elimination of doors, passages between partitions.
- Compatibility of uses:
    - Elimination of intermediate partitions
    - Reducing the surface areas dedicated to service areas

- Integration of:
    - Structural elements in space divisions: load-bearing walls and bracing planes as implicit partitions.
    - Furniture in compartmentalisation with cupboards, shelves, etc.

The rationalised design also facilitates the simplification of superfluous elements, simultaneously reducing costs and timescales.

- Elimination of decorative elements: Skirting boards, ceiling coving, etc.
- Reduction of the cladding of transit spaces: walls and ceilings.
- Elimination of installation aids.
- Elimination of cladding for structural elements (concrete, steel), leaving their exposed qualities.

Rational building design makes it possible to favour cost reduction by reducing the elements of architectural composition.

Rational architecture optimises spaces and construction elements.

These architectural conditions lead to saving in material resources, in the definition of habitat programs, in line with lean methodology.

Later, optimisation of construction can be completed, obtaining better production costs and further reductions in execution times.

As a specific result, applying what is analysed in the article and as an example that materialises industrialisation in housing in an experimental way, one of the authors brought together all the parameters analysed in his Guardiola–Babecka single-family house in Madrid, which has been awarded for its innovative proposal.

Figure 6 shows the interior environment of the house, in which elements of the space have been reduced to define different functions of the habitat.

3.2. Cost Cutting: Resources, Materials and Execution Time

From the construction price decomposition studied in Section 2.4, construction prices are composed of three components:

- Materials.
- Manpower.
- Auxiliary resources.

From these, the following options have been obtained within the framework of cutting and saving time and resources, along the lines of lean construction.

With industrial manufacturing, we reduce the impact of material costs; however, in order to reduce the manpower and auxiliary resources, it is necessary to think about the execution method, simplifying the deadlines.

As far as the construction method is concerned, a mechanised technical assembly allows not only less qualified manpower, with the consequent cost savings (the assemblers could be the users themselves, suitably directed and with assembly instructions), but also independence from climatic conditions, as no materials that have to set in dry weather, are involved.

The repercussion of manpower and auxiliary resources is directly related to the time used for assembly; if the period decreases, these cost components will decrease.

On the other hand, when comparing the prices of traditional materials with those of industrial origin, the price of industrial components shows a downward trend.

Figure 6. Interior of Guardiola–Babecka single-family house in Madrid.

They also serve to guide mechanised construction using the contents of "open indus-trialisation" applied to construction. In summary, the following arguments are the key to sustainable construction.

- Materials can also evolve in an increasingly cost-optimised way.
- Break the dependence of the material on manpower, reducing the impact of the former.
- Lower the impact of auxiliary resources.
- More optimised deadlines: overlapping tasks and obtaining shorter deadlines, delays due to waiting times between trades are reduced or eliminated altogether.
- Avoid drying and setting times for hydraulic materials.

3.3. Streamlined Design: Transfer to Implementation (Productivity Improvement)

A revised design over implementation, through industrialised construction, will improve the evaluation of the final cost.

Based on the above analysis, three strategies for change are observed, for the optimi-sation of the cost of implementation.

3.3.1. Decrease the Impact of Manpower on the Materials

The trend of the material cost for industrial products is downward, in contrast to the manpower, which is upward.

Consequently, a low manpower impact on the price of the material and a material of industrial origin with a negative trend: decreasing cost, generate a doubly decreasing total cost. (Figure 7).

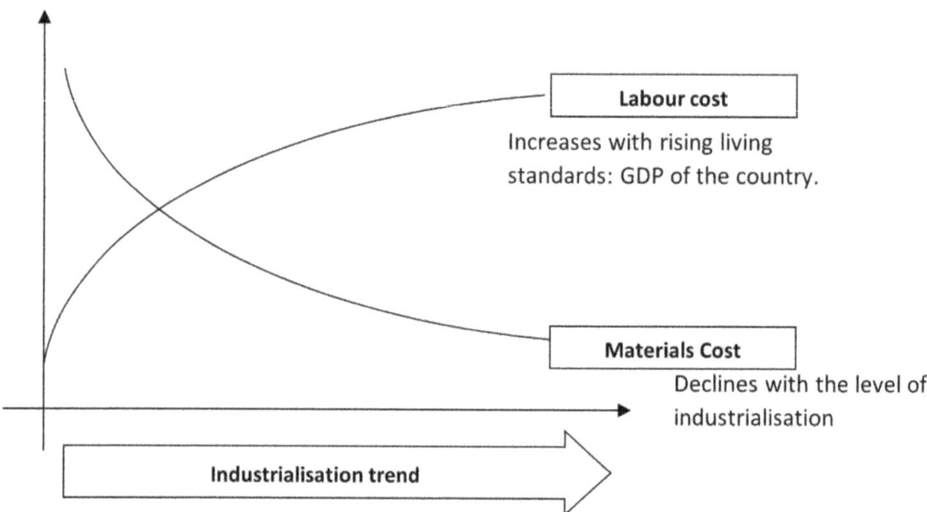

Figure 7. Industrialisation trend (own source).

Two aspects are therefore of interest:

- Materials can also evolve in a cost-optimised manner.
- Break the dependence of material on manpower by reducing the impact of the former.

3.3.2. Lower Impact of Auxiliary Resources: Lower Indirect Costs

Execution based on a design with a low impact of auxiliary resources generates a decreasing cost for two reasons: the elimination of auxiliary resources with a direct impact and the shorter time charged for those of external implementation.

3.3.3. Shortening the Deadline by Shortened Times

Time cuts should be sought in activities that overlap with finished products [33].

- More optimised lead times: with overlapping tasks and shorter lead times, delays due to waiting times between trades are reduced or eliminated altogether.
- Avoid drying and setting times of hydraulic materials.
- Performance improvement towards lean process technology in construction.
- Decrease the impact of manpower.
- Cost reductions due to elimination/cutting of indirect costs.
- Shortening the time limit for reduced times.

In summary, the improvement of performance towards a lean process technology in construction would be stated with the following arguments:

- Decrease the impact of manpower.
- Reduction of costs by eliminating indirect costs: auxiliary resources.
- Shortening of time due to reduced times.

Grouping into the three defined categories of the price under construction decomposition:

- Decrease the impact of manpower on materials.
- Lower repercussion of auxiliary resources: lower indirect cost.
- Shortening of the timeframe due to reduced times.

The prioritisation of material costs over manpower, by means of industrialised components specialised in their function and prepared for rapid assembly on site is characteristic of industrialised construction.

Likewise, the reduction of lead times is an objective of the application of the tried and tested methods of industry in construction.

Both strategies, with the cost of resources and time, add up to the cuts that are characteristic of the application of lean construction.

As a summary of the results obtained in relation to the methodology used, the table in Figure 8 is shown.

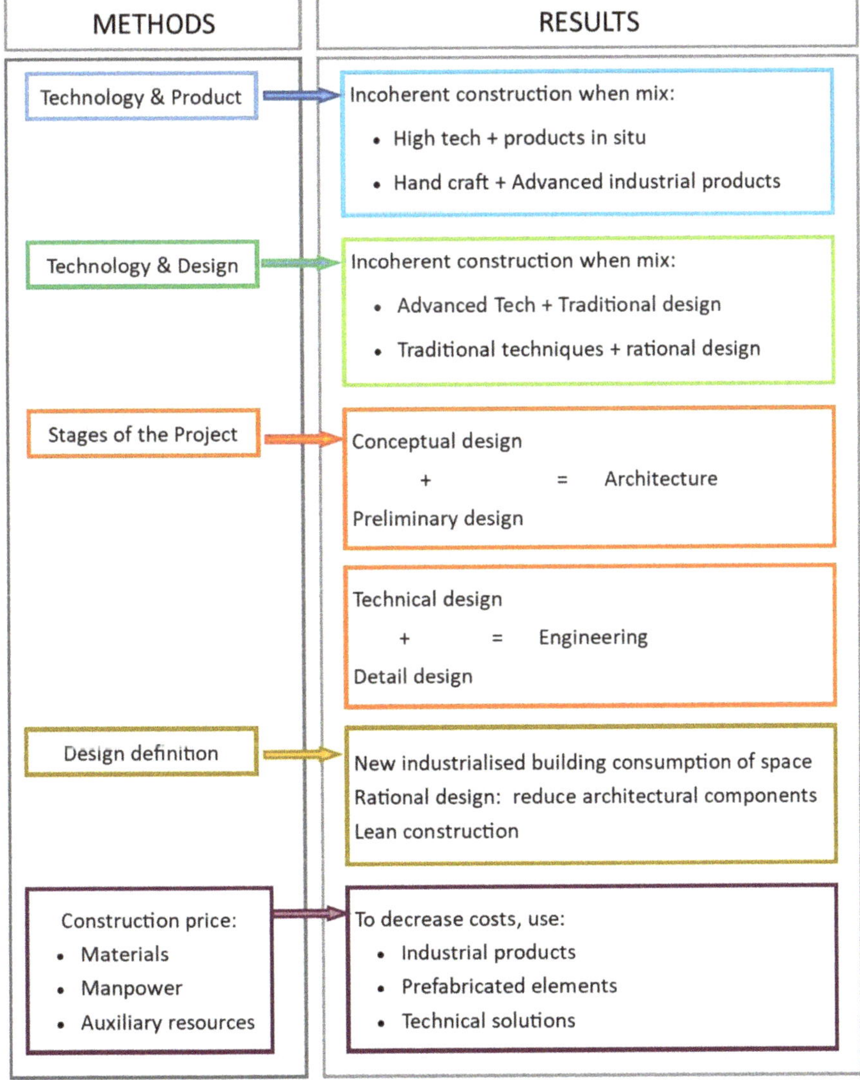

Figure 8. Methodology and results obtained (own source).

4. Discussion

4.1. Coherent Building System

Construction is improved with a complete and coherent building system that integrates: design, project and execution.

As described at the beginning, through the analysis of construction design, integrating design and execution, opportunities have been found to improve productivity in building, justified in lean construction.

In small-scale projects, single-family or block dwellings, small office and commercial buildings, the only technician to develop the design is the architect, whose work usually ends with the basic level of the project, which is specified in the premises of the habitat, developed according to concepts of space organisation, and respecting the legal ordinances and regulations that affect it.

The detailed project for the execution remains in two contributions—the structure and the installations—without going deeper into the constructive analysis, which is taken for granted according to the traditional rules of construction.

In larger projects, the work is complemented by detailed engineering, developed by engineering specialists, outside the architect's organisation, with no other relationship than that of external professional collaboration, normally occasional and without continuity.

However, frequently, even in these cases, either the engineering does not understand the architect's design orientation, or the object of the architect's design is far removed from the parameters of the technical project.

In any case, the execution is not usually developed in-depth, analysing techniques, products and components, which rationalise the construction.

4.2. Economic Parameters

An analysis of the economic parameters was carried out where it was possible to identify cost and time reductions, based on the criteria of the construction itself, without including other disciplines:

- Break the dependence of the material on manpower, reducing the impact of the former.
- Reducing the effect of auxiliary resources.
- More optimized deadlines, overlapping tasks and obtaining shorter deadlines:
    - Delays due to waiting times between trades, which are reduced or eliminated altogether.
    - Avoid drying and setting times for hydraulic materials.
    - Transferring time from site to "workshop".

New architectural optimization arguments have also been included to eliminate components of the habitat programme, which can be grouped into two categories:

1. Simplification of components:

    - Habitat revision.
    - Reduction of interior partitions.
    - Elimination of doors, passages between compartments.
    - Reducing dedicated surfaces in service areas: kitchens, offices.
    - Structural in the divisions of space: load-bearing walls and bracing planes as implicit partitions.

These components define the functional programme of the habitat, which we have seen above, and are basic architectural elements at this level of design, principal elements, and involve both spatial and structural functions.

2. Simplification of superfluous elements:

    - Furniture, in the compartmentalization with cupboards and shelving.
    - Reduction of cladding and transit spaces: walls and ceilings and cladding: structural elements: concrete and steel, leaving their qualities visible.
    - Exposed installations, elimination of installation aids: chases and skirting boards.

On the other hand, these components are of more secondary application for a design level of definition of shapes and textures, more definitions for style than as a functional program.

Performance improvement, towards a lean construction process technology, is summarized:

1. Reductions of architectural elements, for a design, as a solution of the habitat program.
2. Optimization of costs and deadlines in execution:
   - Decrease the impact of manpower on materials.
   - Lower repercussion of auxiliary means: lower indirect cost.
   - Shortening of time due to reduced times.

However, it would be necessary to quantify all these strategic lines, both those of an architectural nature and those of an economic nature, according to the cost and time of construction.

It would also be necessary to obtain a design basis combining the two strategies mentioned, a new method that would allow, in a systemic way, to align with these proposals and which justify a cost reduction on the final construction.

It is difficult to determine the reduction; it depends on the solution adopted, and it is not possible to give a general reduction parameter.

4.3. 3C System: Compatible Components in Construction

As a possible continuation of this article, we propose a design method that allows, within the discipline of design, to integrate the contemporary building into the current industrial environment.

This methodology is developed within systems engineering, adopting concepts and laws from its epistemological field.

Successful implementation of the lean concept as a sustainable approach in the construction industry depends on the identification of critical factors [34].

A system that focuses on industrialized components to obtain the two properties that we have obtained from the previous analysis. (Figure 9).

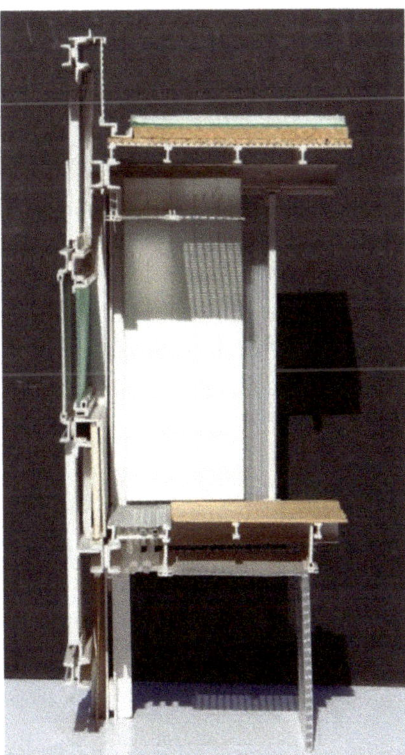

Figure 9. 3C system composition mock-up. J Guardiola A.

To this end, we will seek application in the discipline of systems theory, specifically systems engineering, to find design arguments that achieve the objectives along the lines outlined above.

The model shows how strategic implementation facilitates the integration of lean construction and sustainable construction from the design phase to the finishing phase of a project [35].

A system that we call the 3c system: Compatible Components in Construction.

This proposal brings together these improvement objectives, using the criteria of sustainable construction through the so-called "3C system: Compatible Construction Components". This system involves the use of prefabricated modules, factory materials and assembly using standardised joints. (Figure 10).

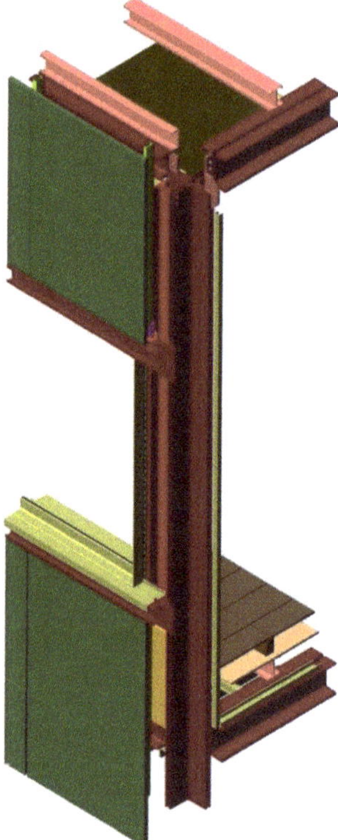

Figure 10. CAD model of the façade component of 3C system composition. J Guardiola A.

A design solution that will seek, first, to formalise its application principles, and then to contrast with traditional solutions, its optimisation operability, as a defined tool of lean construction [36]. (Figure 11).

4.4. Lean Construction

Finally, the lean construction methodology is confirmed as a tool that contributes to the sustainability of buildings [37] for the following reasons based on ISO 20887 [38]. This standard specifies aspects of sustainability in the field of recycling, dismantling and reuse in the field of construction.

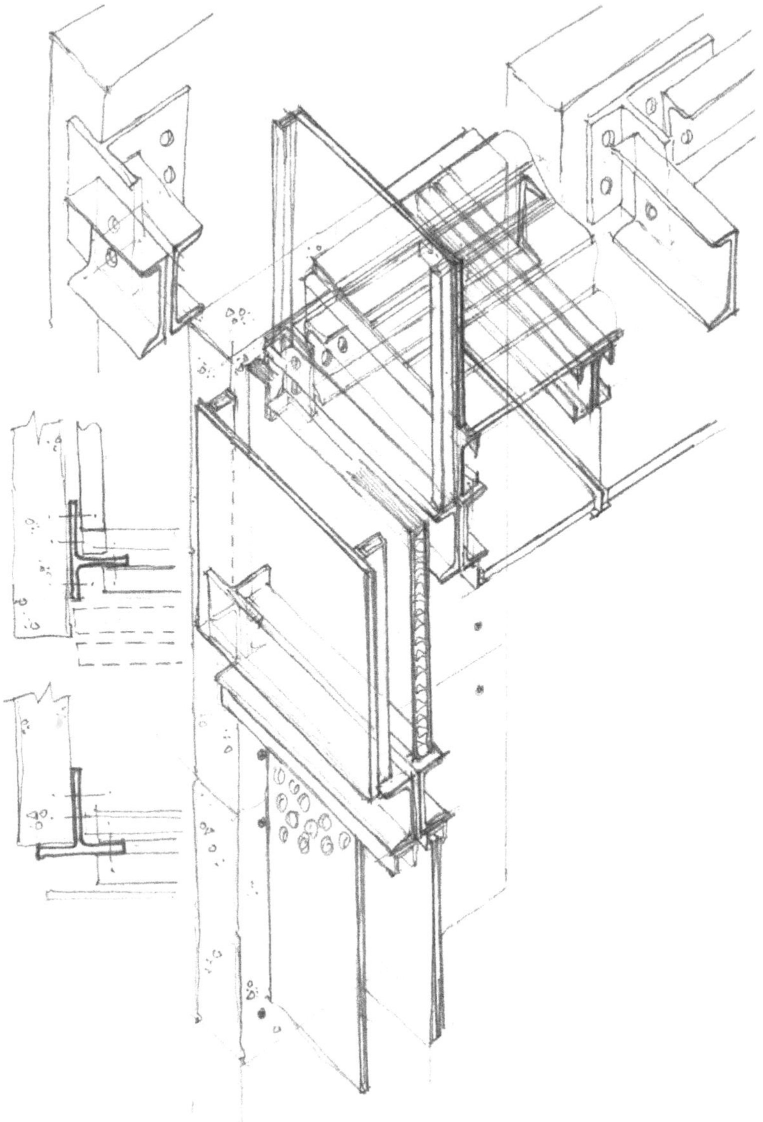

Figure 11. Drawing of the façade component of 3C system composition and connections to other components. J Guardiola A.

The following arguments stand out for the sustainability of the construction:

- For energy savings: on the construction site, the auxiliary resources are responsible for almost the entire energy consumption. This is discussed in the article with regard to the cost element, that of auxiliary resources.
- On materials, their manufacture (consumption), composition (non-polluting), recycling, proximity (transport savings, etc.) and specifically the focus on the "specialisation" of materials. This is also dealt with under the heading of "cost elements".

- Cyclical economy: buildings, components and materials must be used for continuous reuse.

The first two are included in the analysis of cost elements, carried out previously, the third one is considered below, using the above-mentioned ISO 20887 standard as a reference.

This standard, ISO 20887: Design for disassembly and adaptability of buildings and civil engineering works, is structured on specific contents that determine design aspects to be applied to construction assets, of which we highlight the following.

Lean practices show the reduction of environmental, economic, and social impacts during the construction phase, and increase sustainability parameters in the development of projects [39].

Standardisation (seiketsu standardize) facilitates transport, storage, and reuse. It makes it possible to include the aspects of adaptability and simplicity.

Standardisation avoids waste, adapts to logistical, ergonomic, and functional requirements.

The constructive design is favoured by standardisation aspects such as: components, dimensions, and modules.

Simplicity, as a design principle, reduces the number of elements, components and materials to the minimum required to perform the intended function.

This includes limiting the use of decorative details, minimising the number and diversity of materials.

One of the objectives is the elimination of waste in disassembly, aided by the simplification of tools and techniques required for disassembly. This directly improves the appearance of cleanliness (seiso shine), eliminating debris and debris and keeping the workspace tidy.

Lean methodology seeks the reduction of "waste". This "waste" can also be assumed in construction in different categories [40]:

1. Defects
2. Overproduction
3. Waiting
4. Non-utilized skills/capabilities
5. Transportation
6. Inventory
7. Motion
8. Over-processing

By increasing the functions that material provides, it contributes to dematerialisation in construction, thus reducing the problem of material layering.

Order and systematisation in construction is a fundamental aspect of the work. Seiton, set in order, organises, and places the necessary materials according to the order of use by zones on the construction site and establishes the areas where these stockpiles, auxiliary resources, reusable material, and rubble, etc. are to be made.

Independence is particularly important for the "levels" of the construction, related to the temporal prediction of the function of each component.

The main levels that can be identified are:

- Shell or carcass: structure, foundations, and envelopes.
- Services: installations, ducts, pipelines, and machinery.
- Finishes: flooring, ceilings, walls, and panelling.
- Equipment: furniture and lamps.

The challenge is to achieve independence without compromising the functionality of systems and materials in their integration.

Independence facilitates the separation and disconnection of building systems, favouring spatial and functional adaptation and the reuse of systems. In addition, disassembly shall allow materials and components to be removed without disrupting other components or materials.

The levels of a building shall be differentiated to facilitate adaptation and dismantling. Separating levels of long-life components from those of shorter life facilitates retrofitting and reduces the complexity of dismantling.

Durability is defined for an asset or component as the ability to perform as required for its specified period of time without the need for repair or modification.

Durability and adaptability are close concepts.

To minimise maintenance or repair, materials with a long service life should be used. In other cases, shorter life designs are required, with easy disassembly and consequent re-use of components and materials.

The accessibility of components, especially connecting elements with a shorter service life, facilitates replacement and avoids damage to adjacent elements of the assembly. Another advantage is that disassembly of these components from the building sometimes does not require special equipment.

The requirement for easy access to connectors can have an aesthetic impact on the building design.

The discipline of improvement is necessary; shitsuke sustain propose maintenance, sustaining process to create the habit, explaining to the team how it works and how it is maintained.

An open construction, where parts are interchangeable, allows for modifications in the implementation without significant repercussions.

All in all, these reasons contribute to the fact that "Lean Construction not only contributes to creating the economic value to the construction process but can also contribute to promoting the environmental and social issues" [41] and meets several points of the objectives for sustainable development (OSD):

With the development of new building techniques [10], lower consumption of material and energy resources and the planning of the building life cycle, more sustainable cities and population centres can be achieved [12].

Controlling resources and reducing waste through Lean techniques will lead to responsible production and consumption [13].

5. Conclusions

In order to improve productivity, in the current situation of construction, the transfer to industrialisation is essential.

With a complete building system, integrated by design, project and execution, the construction is improved and more sustainable.

It can be determined that incoherent construction occurs when mixing craft technologies with factory-produced materials or industrial technologies using materials made on site. (Figure 1).

Incoherent construction is also made when advanced technologies are used with a traditional design or when traditional techniques are combined with a rational design. (Figure 2).

A lean analysis in construction must be based on the order of stages established in construction projects, as set out in the UNE-EN 15643-3-2012. Of the four design stages, conceptual design and preliminary design would be in the project architecture group, while technical design and detail design would be in the execution engineering group.

The habitat system must correspond to new modes of construction, which respect the arguments obtained by the new type of buildings.

Reducing the cost of construction is of great importance, both from an economic and a social point of view.

Attempts to reduce the cost of traditional manual labour methods by introducing more rigorous organisational techniques have so far produced only a little progress.

On the other hand, numerous studies, both technical and sociological, have reviewed the human habitat, looking for more rational solutions, which have an impact on space-saving.

The repercussion of manpower and auxiliary resources is directly related to the time used for assembly; if the period of time decreases, these cost components will decrease.

The new objective, uniting both trends, is to seek construction by means of industrial production methods with spaces that meet the real needs of the habitat.

Industrial production must necessarily go through a manufacturing process at the ground level, using components that can be assembled "in situ". Modern technology is already prepared to take this step, but today's construction still uses archaic manual methods, in which the machine plays only an auxiliary role.

From the project phases of architectural design, reviewing the habitat and the defining components of the spaces to the execution phases of engineering and analysing process improvements from the analytical decomposition of the prices of building components, opportunities have been found to improve the construction process.

A new methodology should bring together these improvement objectives, using the criteria of construction sustainability, through the so-called "3C system: Compatible Building Components". This system involves the use of prefabricated modules, factory materials and assembly by means of standardised joints.

The application of the BIM methodology in the building's life cycle can go from the beginning of the design to its demolition. This building life can be programmed efficiently, if, during the design process and its execution, the lean methodology is applied, seeking to reduce waste, saving time and materials.

As a final conclusion, it is established that the elements that must integrate a complete and coherent building project require the incorporation of industrial manufacturing and lean methodology in the construction processes in order to achieve sustainable architecture.

Author Contributions: Conceptualization, T.A. and J.G.; methodology, T.A. and J.G.; software, D.F.; validation, T.A. and D.F.; formal analysis, T.A. and J.G.; investigation, T.A. and J.G.; resources, J.G.; data curation, T.A. and D.F.; writing—original draft preparation, J.G. and T.A.; writing—review and editing, T.A.; visualization, T.A. and D.F.; supervision, J.G. and D.F.; project administration, T.A. and D.F.; funding acquisition, T.A., J.G. and D.F. All authors have read and agreed to the published version of the manuscript.

Funding: This research received no external funding.

Institutional Review Board Statement: Not applicable.

Informed Consent Statement: Not applicable.

Acknowledgments: Thanks to all the organizations and people who participated in the project, particularly to professors who have a spirit of continuous improvement.

Conflicts of Interest: The authors declare no conflict of interest. The funders had no role in the design of the study; in the collection, analyses, or interpretation of data; in the writing of the manuscript, or in the decision to publish the results.

References

1. Resta, B.; Dotti, S.; Gaiardelli, P.; Boffelli, A. Lean Manufacturing and Sustainability: An Integrated View. In *Proceedings of the IFIP International Conference on Advances in Production Management Systems, Iguassu Falls, Brazil, 3–7 September 2016*; Springer: Berlin/Heidelberg, Germany, 2016; pp. 659–666.
2. Gao, S.; Low, S.P. The Toyota Way. In *Lean Construction Management: The Toyota Way*; Gao, S., Low, S.P., Eds.; Springer: Singapore, 2014; pp. 49–100, ISBN 978-981-287-014-8.
3. Marhani, M.A.; Jaapar, A.; Bari, N.A.A. Lean Construction: Towards Enhancing Sustainable Construction in Malaysia. *Procedia—Soc. Behav. Sci.* **2012**, *68*, 87–98. [CrossRef]
4. Abdelhamid, T. Lean construction: Where are we and how to proceed. *Lean Constr.* **2004**, *1*, 24.
5. Sandvik, C.; Fougner, F. BIM as a Tool for Sustainable Design. Proceedings of 5th fib Congress, Melbourne, Australia, 8 October 2018.
6. United Nations. *Transforming Our World: The 2030 Agenda for Sustainable Development*; A/RES/70/1; UN: New York, NY, USA, 2015.
7. Bae, J.-W.; Kim, Y.-W. Sustainable Value on Construction Projects and Lean Construction. *J. Green Build.* **2008**, *3*, 156–167. [CrossRef]

8. Goñi, P.M.; Barroso, J.M.G.; Rey, A.R.E. The "Design for Disassembly", a Lean Methodology. In Proceedings of the 5th European Conference on Energy Efficiency and Sustainability in Architecture and Planning, San Sebastian, Spain, 7–9 July 2014. [CrossRef]

9. Saieg, P.; Sotelino, E.D.; Nascimento, D.; Caiado, R.G.G. Interactions of Building Information Modeling, Lean and Sustainability on the Architectural, Engineering and Construction Industry: A Systematic Review. *J. Clean. Prod.* **2018**, *174*, 788–806. [CrossRef]

10. Tzortzopoulos, P.; Kagioglou, M.; Koskela, L. *Lean Construction: Core Concepts and New Frontiers*; Routledge: London, UK, 2020; ISBN 978-0-429-51215-5.

11. Solaimani, S.; Sedighi, M. Toward a Holistic View on Lean Sustainable Construction: A Literature Review. *J. Clean. Prod.* **2020**, *248*, 119213. [CrossRef]

12. Brookfield, E.; Emmitt, S.; Hill, R.; Scaysbrook, S. The Architectural Technologist's Role in Linking Lean Design with Lean Construction. In Proceedings of the 12th Annual Conference of the International Group for Lean Construction, Helsingør, Denmark, 3–5 August 2004; p. 7.

13. Mathaisel, D.F.X. A Lean Architecture for Transforming the Aerospace Maintenance, Repair and Overhaul (MRO) Enterprise. *Int. J. Product. Perform. Manag.* **2005**, *54*, 623–644. [CrossRef]

14. Ogunbiyi, O.; Goulding, J.S.; Oladapo, A. An Empirical Study of the Impact of Lean Construction Techniques on Sustainable Construction in the UK. *Constr. Innov.* **2014**, *14*, 88–107. [CrossRef]

15. Emuze, F.A.; Saurin, T.A. *Value and Waste in Lean Construction*; Routledge: London, UK, 2015; ISBN 978-1-317-44763-4.

16. Koskela, L. *Application of the New Production Philosophy to Construction*; CIFE Technical Report; Stanford University: Stanford, CA, USA, 1992; Volume 72.

17. Rajadell, M.; Sánchez, J. *Lean Manufacturing: La Evidencia de Una Necesidad*; Díaz de Santos: Madrid, Spain, 2010.

18. Corbusier, L.; Eardley, A. *The Athens Charter*; Grossman Publishers: New York, NY, USA, 1973; ISBN 0-670-13970-X.

19. Fieber, K.; Mills, J.; Peppa, M.V.; Haynes, I.; Turner, S.; Turner, A.; Douglas, M.; Bryan, P. Cultural heritage through time: A case study at Hadrian's wall, United Kingdom. *ISPRS—Int. Arch. Photogramm. Remote Sens. Spat. Inf. Sci.* **2017**, *42*, 297–302. [CrossRef]

20. Koolhaas, R.; Obrist, H.U. *Project Japan, Metabolism Talks*; Taschen: Cologne, Germany, 2011; ISBN 978-3-8365-2508-42019.

21. Sadler, S. *Archigram: Architecture without Architecture*; Mit Press: Cambridge, MA, USA, 2005; ISBN 0-262-69322-4.

22. Garnier, T. *An Industrial City*; Princeton Architectural Press: New York, NY, USA, 1989; ISBN 0-910413-47-9.

23. Oechslin, W.; Wang, W. Les Cinq Points d'une Architecture Nouvelle. *Assemblage* **1987**, *4*, 83–93. [CrossRef]

24. Fernández, J.R. La Arquitectura Prefabricada de Rafael de la Hoz en Córdoba: Entre el Detalle Constructivo y la Generación del Proyecto. Universidad de Sevilla, 2016. Available online: https://dialnet.unirioja.es/servlet/tesis?codigo=47825 (accessed on 22 October 2021).

25. Radić, J.; Kindij, A.; Mandić Ivanković, A. *History of Concrete Application in Development of Concrete and Hybrid Arch Bridges*; Chinese-Croatian Joint Colloquium Long Arch Bridges: Brijuni Island, Croatia, 10–14 July 2008; p. 24.

26. Stanford University and the 1906 Earthquake. Available online: https://quake06.stanford.edu/centennial/gallery/structures/museum/index.html (accessed on 22 October 2021).

27. Association pour la mémoire et le rayonnement des travaux d'Eugène Freyssinet (Ed.) *Eugène Freyssinet: A Revolution in the Art of Construction*; Presses de l'école nationale des ponts et chaussées: Paris, France, 2004; ISBN 978-2-85978-394-5.

28. Smith, R.E. *Prefab Architecture: A Guide to Modular Design and Construction*; John Wiley & Sons: Hoboken, NJ, USA, 2010; ISBN 978-0-470-27561-0.

29. Moles, A.A. *Rohmer, Elisabeth Sicología del Espacio*; Editorial Ricardo Aguilera: Madrid, Spain, 1972; ISBN 978-84-7005-114-2.

30. Fernández Alba, A. *La Crisis de la Arquitectura Española (1939–1972)*; E.T.S. Arquitectura (UPM): Madrid, Spain, 1972.

31. Díaz, H.P.; Rivera, O.G.S.; Guerra, J.A.G. Filosofía Lean Construction para la gestión de proyectos de construcción. *Avances Investigacion en Ingeniería* **2014**, *11*, 32–53. [CrossRef]

32. Kwofie, S.; Pasquire, C. Lean Thinking for Structural Engineers. In Proceedings of the 28th Annual Conference of the International Group for Lean Construction (IGLC), Berkeley, CA, USA, 6–10 July 2020.

33. Aslam, M.; Gao, Z.; Smith, G. Exploring Factors for Implementing Lean Construction for Rapid Initial Successes in Construction. *J. Clean. Prod.* **2020**, *277*, 123295. [CrossRef]

34. Dehdasht, G.; Ferwati, M.S.; Zin, R.M.; Abidin, N.Z. A Hybrid Approach Using Entropy and TOPSIS to Select Key Drivers for a Successful and Sustainable Lean Construction Implementation. *PLoS ONE* **2020**, *15*, e0228746. [CrossRef] [PubMed]

35. Jamil, A.H.A.; Fathi, M.S. The Integration of Lean Construction and Sustainable Construction: A Stakeholder Perspective in Analyzing Sustainable Lean Construction Strategies in Malaysia. *Procedia Comput. Sci.* **2016**, *100*, 634–643. [CrossRef]

36. Carvalho, B.S.D.; Scheer, S. Lean as an Integrator of Modular Construction. *Modul. Offsite Constr. (MOC) Summit Proc.* **2019**, 149–156. [CrossRef]

37. Ogunbiyi, O.E. Implementation of the Lean Approach in Sustainable Construction: A Conceptual Framework. Ph.D. Thesis, University of Central Lancashire, Lancashire, UK, 2014.

38. ISO Comitee ISO 20887:2020 Sustainability in Buildings and Civil Engineering Works—Design for Disassembly and Adaptability—Principles, Requirements and Guidance. Available online: https://www.iso.org/cms/render/live/en/sites/isoorg/contents/data/standard/06/93/69370.html (accessed on 22 October 2021).

39. Carvajal-Arango, D.; Bahamón-Jaramillo, S.; Aristizábal-Monsalve, P.; Vásquez-Hernández, A.; Botero, L.F.B. Relationships between Lean and Sustainable Construction: Positive Impacts of Lean Practices over Sustainability during Construction Phase. *J. Clean. Prod.* **2019**, *234*, 1322–1337. [CrossRef]

40. Tafazzoli, M.; Mousavi, E.; Kermanshachi, S. Opportunities and Challenges of Green-Lean: An Integrated System for Sustainable Construction. *Sustainability* **2020**, *12*, 4460. [CrossRef]
41. Bajjou, M.S.; Chafi, A.; Ennadi, A.; El Hammoumi, M. The Practical Relationships between Lean Construction Tools and Sustainable Development: A Literature Review. *J. Eng. Sci. Technol. Rev.* **2017**, *10*, 170–177. [CrossRef]

MDPI

St. Alban-Anlage 66

4052 Basel

Switzerland

Tel. +41 61 683 77 34

Fax +41 61 302 89 18

www.mdpi.com

Sustainability Editorial Office

E-mail: sustainability@mdpi.com

www.mdpi.com/journal/sustainability